Industrial Instrumentation

Industrial Instrumentation

Dr. Al Sutko
Dr. Jerry D. Faulk
Midwestern State University

Delmar Publishers

I(T)P An International Thomson Publishing Company

Albany • Bonn • Boston • Cincinnati • Detroit • London • Madrid
Melbourne • Mexico City • New York • Pacific Grove • Paris
San Francisco • Singapore • Tokyo • Toronto • Washington

NOTICE TO THE READER

Publisher does not warrant or guarantee any of the products described herein or perform any independent analysis in connection with any of the product information contained herein. Publisher does not assume, and expressly disclaims, any obligation to obtain and include information other than that provided to it by the manufacturer.

The reader is expressly warned to consider and adopt all safety precautions that might be indicated by the activities herein and to avoid all potential hazards. By following the instructions contained herein, the reader willingly assumes all risks in connection with such instructions.

The publisher makes no representation or warranties of any kind, including but not limited to, the warranties of fitness for particular purpose or merchantability, nor are any such representations implied with respect to the material set forth herein, and the publisher takes no responsibility with respect to such material. The publisher shall not be liable for any special, consequential, or exemplary damages resulting, in whole or part, from the readers' use of, or reliance upon, this material.

Cover Design: Aurora Design

Delmar Staff
Publisher: Robert Lynch
Administrative Editor: John Anderson
Editorial Assistant: John Fisher
Senior Project Editor: Christopher Chien
Production Manager: Larry Main
Art and Design Coordinator: Nicole Reamer

COPYRIGHT © 1996
By Delmar Publishers
a division of International Thomson Publishing Inc.

The ITP logo is a trademark under license.

Printed in the United States of America

For more information, contact:

Delmar Publishers
3 Columbia Circle, Box 15015
Albany, New York 12212-5015

International Thomson Publishing Europe
Berkshire House 168-173
High Holborn
London, WC1V 7AA
England

Thomas Nelson Australia
102 Dodds Street
South Melbourne, 3205
Victoria, Australia

Nelson Canada
1120 Birchmont Road
Scarborough, Ontario
Canada, M1K 5G4

International Thomson Editores
Campos Eliseos 385, Piso 7
Col Polanco
11560 Mexico D F Mexico

International Thomson Publishing GmbH
Konigswinterer Strasse 418
53227 Bonn
Germany

International Thomson Publishing Asia
221 Henderson Road
#05-10 Henderson Building
Singapore 0315

International Thomson Publishing—Japan
Hirakawacho Kyowa Building, 3F
2-2-1 Hirakawacho
Chiyoda-ku, Tokyo 102
Japan

All rights reserved. No part of this work covered by the copyright hereon may be reproduced or used in any form or by any means—graphic, electronic, or mechanical, including photocopying, recording, taping, or information storage and retrieval systems—without the written permission of the publisher.

1 2 3 4 5 6 7 8 9 10 XXX 01 00 99 98 97 96

Library of Congress Cataloging-in-Publication Data

Faulk, Jerry.
   Industrial instrumentation / Jerry Faulk, Adolph A. Sutko.
     p. cm.
   Includes index.
   ISBN 0-8273-6125-4
   1. Industrial electronics. 2. Electronic instruments. I. Sutko,
Adolph A. II. Title.
TK7881.F38 1996
681'.2—dc20
                                    94-2656
                                       CIP

Contents

	Preface	ix
1	**Introduction** Read By 8-28	1
	Basic Considerations / **2**	
	Process Control / **2**	
	Static Characteristics of Instruments / **8**	
	Review Materials / **15**	
2	**DC Electricity**	19
	Basic Considerations / **20**	
	Direct Current / **20**	
	Review Materials / **45**	
3	**AC Electricity**	51
	Basic Considerations / **52**	
	Alternating Current / **53**	
	Review Materials / **75**	
4	**Electronics**	79
	Basic Considerations / **80**	
	Analog Electronics / **80**	
	Digital Electronics / **90**	
	Review Materials / **92**	

5 Pressure — 95
Basic Considerations / 96
Measuring Devices / 105
Application Considerations / 117
Review Materials / 122

6 Signal Transmission — 129
Electrical / 130
Pneumatic / 141
Fiber Optics / 142
Review Materials / 143

7 Temperature and Heat — 147
Basic Considerations / 148
Measuring Devices / 159
Application Considerations / 168
Review Materials / 172

8 Level — 177
Basic Considerations / 178
Measuring Devices / 179
Application Considerations / 190
Review Materials / 194

9 Flow — 197
Basic Considerations / 198
Measuring Devices / 210
Application Considerations / 221
Review Materials / 223

10 Humidity — 229
Basic Considerations / 230
Measuring Devices / 235
Application Considerations / 240
Review Materials / 241

11 Other Variables — 243
  Density and Specific Gravity / **244**
  Viscosity / **249**
  Position/Displacement / **254**
  Force, Torque, and Load Cells / **255**
  Sound / **260**
  pH Measurements / **265**
  Review Materials / **267**

12 Process Control — 273
  Controller Action / **274**
  Implementation of Controller Action / **291**
  Review Materials / **300**

Appendix A Instrument Society of America (ISA) Symbols — 305
Appendix B Answers to Odd-Numbered Problems — 309
  Glossary — 319
  Index — 329

Delmar Publishers' Online Services

To access Delmar on the World Wide Web, point your browser to:

http://www.delmar.com/delmar.html

To access through Gopher: gopher://gopher.delmar.com
(Delmar Online is part of "thomson.com", an Internet site
with information on more than 30 publishers
of the International Thomson Publishing organization.)

For information on our products and services:
email: info@delmar.com
or call:
800-347-7707

Preface

This text is for those who are interested in instrumentation used in industry today. Its primary users will be students in two- or four-year engineering technology programs. Others that will find it useful are engineering students, students in technical or vocational schools, and anyone who is interested in how and why measuring instruments work. People who are or will be involved in such areas as planning, designing, operating, testing, analyzing, evaluating, or maintaining equipment will find answers to questions that arise as they do their jobs.

Although the topics covered in this text come from a broad area, including engineering, physics, chemistry, and electronics, it is not necessary that the user be knowledgeable in these topics. The approach is primarily from a practical point of view with enough theory included to answer the basic questions that sometimes arise.

In general, each chapter contains a discussion of basic terms and concepts associated with the subject of that chapter, a description of several types of possible measuring devices, and concludes with considerations involved with applications. Chapter 1 gives an introduction to the subject of industrial instrumentation. Chapters 2, 3, and 4 cover the areas of electricity and electronics as they will be used throughout the book. Chapters 5, 7, 8, 9, 10, and 11 deal with pressure, temperature and heat, level, flow, humidity, and other pertinent variables, respectively. Chapter 6 provides information on signal transmission with respect to measurements being made. The text concludes with Chapter 12, which deals with the broad area of process control. Although it is not the purpose of this text to deal in depth with the different systems of units in use today, the use of other-than-engineering units is given in several instances.

The authors' primary reason for writing this book was that they were faced with teaching a course in basic instrumentation and could not find a recently written text that covered the necessary topics and that had enough examples in the body of the chapters and questions and problems at the ends of the chapters. It was also their goal to keep the book relatively short, since, in their opinion, excessive length was one of the drawbacks of the older texts available. Because the authors have been involved with industrial activities for many years, some of the text may be slanted toward their experiences and interests. In general, the sources for the material are experience from industry and books, journals, and magazines.

Use of the text is straightforward, with each of the chapters essentially standing by itself. Some chapters have equations used in the developments; but, in most cases, it would be entirely appropriate to bypass the development equations and deal only with the application formulas or equations. Material marked with a vertical line in the margin can be omitted in low-level or time-restricted courses. All chapters are followed by questions and problems that deal with the chapter topic. An instructor's guide that includes answers for both the questions and problems is available.

The authors would like to dedicate the text to past, present, and future manufacturing engineering technology students here at Midwestern State University. Their comments, questions, suggestions, and needs have been a driving force. Finally, many thanks to Jacque Williams, manufacturing engineering technology secretary, for her work in all aspects of the manuscript preparation.

Introduction

CHAPTER GOALS

After completing study of this chapter, you should be able to do the following:

Understand why the subject of instrumentation is vitally important to the industrial world.

Explain what process control is.

Know the difference between a manipulated variable and a controlled variable.

Discuss the major components of a process control system.

List the major static characteristics of instruments.

Calculate any static characteristic of an instrument, given the pertinent data.

Discuss the difference between an analog instrument and a digital instrument.

Instrumentation is one of the backbone elements of the industrial environment. It is present everywhere in industry, ranging in form from simple pressure gages to sophisticated laser interferometers. Instrumentation is used to control the thermostat for the heating/air conditioning of the company office. It is used to measure the level of the raw materials in the hoppers and supply bins at the heads of production lines. Without instrumentation, industry could not begin to succeed. This book is about instrumentation. It is about the types of instruments available and how they work. It is about applying these instruments to simple uses and to complex uses.

BASIC CONSIDERATIONS

The subject of instrumentation means different things to different people in the world of technology. One can pick up a book with a title containing the subject and find that it covers nothing but electronic test instruments: voltmeters, ohmmeters, multimeters, oscilloscopes, and the like. Instrumentation is much more than simply electronic devices, however. The subject includes electrical devices (both analog and digital), pneumatic devices, hydraulic devices, mechanical devices, and some devices that are so simple one would never think of them as instruments. A ruler, for example, provides measurement of a variable; hence, it is an instrument. An *instrument* is a device used to measure a physical variable. Oftentimes, the instrument will also display and record the value of the variable.

Many different types of physical variables can be measured by instruments. In the case of the ruler, the variable is length. Among the physical variables covered in this textbook are pressure, current, voltage, temperature, level, flow, and humidity. These are all very important factors in the industrial environment.

PROCESS CONTROL

One of the most important uses of instrumentation is in the field of process control. A *process* is a sequence of operations carried out to achieve a desired end result. Fixing a flat tire is a process. Having a hot water heater provide water for an industrial bath several feet away from the heater and, at the same time, vary the temperature of the water so that the water bath remains at a constant temperature regardless of any changes going on in the bath, is a process. There is a big difference between these two processes, however. The first one is manual; the second is automatic. Automatic process control is one of the main subjects of this textbook. It is more commonly referred to simply as *process control*.

Process control is the automatic holding of certain process variables within given limits. This is done by the following steps (simplified for this introduction): (1) a sensor detects the values of the variables associated with the output of the process, (2) these values are sent to a device known as a controller, (3) the controller compares the values of the variables to what their values should be, (4) the controller sends corrective signals to an actuator, and (5) the actuator changes the input variables of the process. The five steps are then repeated; hence, a loop exists. Process control is always in loop form because it is self-correcting.

A simplified view of process control is shown in Figure 1-1. This is a manual process control system, of course. If the person were replaced by an inanimate controller, one would have automatic process control in action.

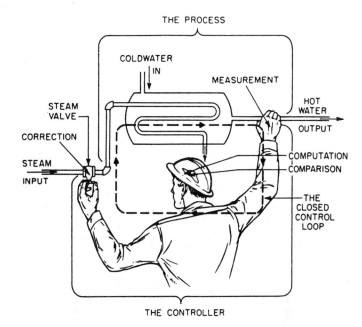

Figure 1-1 Example of manual process control *(Courtesy of D. Considine,* Process Instruments and Controls, *3rd ed., © 1985, McGraw-Hill, Inc. Reprinted with permission of McGraw-Hill, Inc.)*

The sensor in Figure 1-1 is the person's right hand. The controller is his brain. The actuator is the combination of left hand and valve. The process is maintaining a flow of water at a constant temperature. A simplified schematic of this is shown in Figure 1-2.

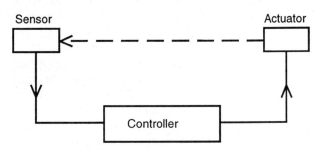

Figure 1-2 Very simple block diagram of process control.

4 / INTRODUCTION

Figure 1-3 Basic block diagram of process control.

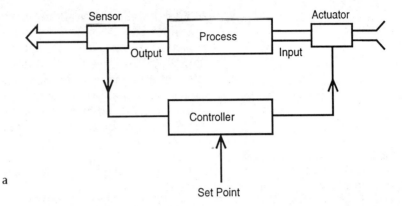

In actuality, this schematic is too simple since it does not indicate the involvement of a process. A better diagram is shown in Figure 1-3a, and an even more detailed diagram in Figure 1-3b. Here we have two variables being modified or measured, separated by the process. The actuator changes one of the variables involved in the process, and the sensor/transducer measures the variable associated with the output of the process. The variable manipulated by the actuator is known as the *manipulated variable*. The variable measured to indicate the condition of the output of the process is known as the *measured variable* or *controlled variable*. Some confusion can be caused by the term *controlled variable* since the variable manipulated by the actuator is also a controlled variable. The best way to determine the controlled variable for any given process is to ask oneself, "What variable was this process control system set up to control?" The answer to this question, then, will also be the answer to the question, "What is the controlled variable for this process?"

The same problem can arise with the term *manipulated variable*, since both the input and the output to the process have been manipulated. The manipulated variable, however, is always the one that is directly manipulated by the actuator. This should be committed to memory.

The terms *sensor, transducer, actuator,* and *controller* were mentioned earlier and now need to be defined. They are all devices and are all part of the process control loop. Because of this, they are referred to as *elements*.

A *sensor* is a device that senses one of the fundamental physical variables, such as temperature, humidity, or pressure. Sensing the variable means that the device can detect not only that the variable is present but also to what degree it is present. A human finger is a sensor because it can tell if heat energy is present and, to a very inaccurate extent, how much heat energy is present. A thermometer is another sensor that can determine if heat energy is present (actually, some heat energy is always present), but it can determine to what degree it is present much more accurately than the human finger. (The measure of heat energy present is most commonly called *temperature*.)

A *transducer* is a device that changes one form of energy to another. For example, a device known as a thermocouple senses heat energy and changes it to an output of electrical energy. An I/P transducer senses current (I) and outputs a proportional amount of pressure (P). Sensors are frequently an integral, built-in part of a transducer and henceforth the combined sensor/transducer will often be referred to simply as a sensor.

An *actuator* is a device that performs an action on one of the input variables of a process according to a signal received from the controller. The actuator would probably be a valve in the case of a heat exchange system. Other examples are a motor speed control where the motor is driving a conveyor belt, and the magnetic relays that turn on and off the fans and compressor in a central air conditioning system.

The *controller* is the element of the process control loop that decides what action needs to be taken. It is the "mind" of the loop. The controller compares the incoming signal from the sensor with the set point and, if there is a difference between the two, sends a corrective signal to the actuator. The *set point*, a fixed signal fed into the controller, has a value equal to the signal that would be sent from the sensor if the measured variable were at its desired value.

Another device that should be mentioned at this point is a transmitter. A *transmitter* conditions the signal received from a transducer so that it is strong enough to be sent some distance away to a receiver such as a controller. For example, the output of a thermocouple is electrical voltage; but it is too weak to be sent over any appreciable length of wire before it is simply lost. Instead the output of the thermocouple typically travels over only two to three feet of wire to a transmitter. The transmitter takes the microvolt output of the thermocouple and amplifies it into, perhaps, a range of 1 to 10 volts. This signal can safely and accurately be sent over several feet of transmission cable, oftentimes a few hundred feet.

Now that several of the terms involved in process control have been discussed, a more detailed process control loop can be drawn, such as shown in Figure 1-4.

In Figure 1-4, the process is the controlled supply of hot water to be maintained at a fixed temperature. The transducer senses the temperature of the hot water leaving the heat exchanger and sends a weak signal to the transmitter. The transmitter amplifies and conditions the signal and sends

Figure 1-4 Detailed block diagram of an example of process control.

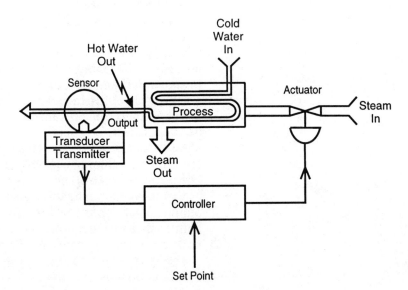

it to the controller. The controller compares the incoming signal to the set point. The set point value is equal to the strength of the signal supplied by the transmitter when the hot water leaving the heat exchanger is at the sought-after temperature. The controller sends an output signal to the actuator that is, in most cases, proportional to the difference between the incoming signal and the set point. The actuator then manipulates the amount of steam passing through the heat exchanger according to the strength of the signal received from the controller.

For example, suppose the aforementioned system were set up to maintain hot water at 200° F. Further suppose that the water leaving the exchanger was at 190° F. The transducer would sense a temperature of 190° F and provide a signal representing this temperature to the transmitter, which would amplify the signal and send it to the controller. As an example, say the signal arrives as a 4.5 v voltage signal. If the system were using a transmission range of 0 to 10 v for its signals, then a very reasonable set-point signal for representing the ideal temperature of 200° F would be 5 volts. In the present case, then, the controller would compare the 4.5 v incoming signal with the 5 v set point and then send an output signal to the actuator proportional to the difference between the 4.5 v and the 5 v. The actuator would be preadjusted so that this arriving signal from the controller would be enough to cause the actuator to open up and send enough additional steam through the heat exchanger to raise the temperature of the hot water output to 200° F.

EXAMPLE 1-1

Shown in Figure 1-5 is a diagram of a process control system for a drying oven. Identify the following elements: (a) the controller, (b) the actuator, (c) the sensor. Also, identify the following: (d) the measured variable, (e) the controlled variable, (f) the manipulated variable, (g) the process.

(a) The controller is the block labelled "controller." (b) The motor speed control. It would also be correct to denote the motor and the motor speed control combination as the actuator. (c) The photocell. (d) The dryness of the objects that were passed through the oven or, more directly, the darkness of the objects or the amount of light reflected by the objects. (e) Same as d. (f) The speed of the conveyor belt. (g) Drying to a controlled degree of dryness.

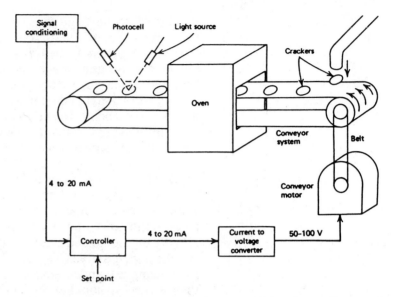

Figure 1-5 A process control example of a drying oven *(From Curtis Johnson,* Process Control Instrumentation Technology, *4th ed., © 1993, p. 275. Reprinted by permission of Prentice Hall, Englewood Cliffs, N.J.)*

Thus far, ideal process control systems have been discussed. In actuality many more problems need to be solved in order to produce a workable process control system. These problems include overshoot, time lag, long-term load changes, and many others. These are dealt with in the chapter on process control. Before advancing to that chapter, however, many preliminary subjects must be covered, especially the instruments used for the sensors and transducers and the methods of signal transmission.

STATIC CHARACTERISTICS OF INSTRUMENTS

Instruments are characterized by several different factors. These factors specify the accuracy of the instrument in question, the precision, the linearity, the reproducibility, the repeatability, and several other characteristics. The value of these factors can oftentimes be found on the specifications sheet for the instrument. These characteristics can pertain to the performance of the instrument in a static or a very slowly changing situation, such as a pressure gage on an air compressor tank. They can also refer to a dynamic situation, such as the case of a pressure gage on a line where the pressure is rapidly changing. Of the two characteristic categories, only the static situation is appropriate for the level of this book. Within the static characteristic category, only the three most

Figure 1-6 Analog display
(Courtesy of the L. S. Starrett Co.)

Figure 1-7 Digital display *(Courtesy of John Fluke Manufacturing Co., Inc. Reproduced with permission)*

widely used factors will be discussed in detail: (1) accuracy, (2) precision, and (3) reproducibility.

Discussion of the first two of these characteristics is also divided into two parts: (1) digital instruments and (2) analog instruments. Digital instruments typically have a display that can give values that are only *discretely* different from one another. Analog instruments have a *continuous* display. An example of an analog display is shown in Figure 1-6. A typical digital display would be that of a digital multimeter as shown in Figure 1-7.

Accuracy

The *accuracy* of an instrument is a measure of the difference between the indicated values given by the instrument and the true values. Accuracy is determined by comparing an indicated value to an accepted standard value. This standard value will be a calibration standard whose ancestry is traceable to the National Institute of Standards and Technology, formerly known as the National Bureau of Standards. The result is nearly always stated in a "±" amount.

Accuracy is most commonly specified as *percentage full-scale accuracy*, abbreviated *%FS*. Percentage full scale is calculated by dividing the accuracy of the instrument by its full-scale output.

EXAMPLE 1-2

If the largest pressure that can be read with a pressure gage is 30 lbs/in² and the gage's accuracy is ±2.5 lbs/in², what is the %FS accuracy?

$$\%FS = \frac{\pm 2.5 \text{ lbs/in}^2}{30 \text{ lbs/in}^2} \times 100\% = \pm 8.33\%$$

The accuracy for a digital instrument would typically be expressed as a percentage of its maximum display amount. For example, if a digital voltmeter has a display with four digits available in the form of XX.XX, then the maximum amount of voltage that can possibly be shown is 99.99 volts. This amount will often be the top of the range for the instrument, but not always. It is not uncommon at all to have a digital instrument with a XX.XX form of display with a maximum amount of 19.99 or 49.99 or another quantity lower than 99.99. Specification sheets usually note the maximum display quantity for a digital instrument.

Another way accuracy is specified is by percentage of the span. The *span* on an instrument is the range between the smallest value the device can display or indicate and the largest value. For example, a pressure gage with a scale that begins at 100 lbs/in^2 and ends at 1,000 lbs/in^2 would have a span of 100 to 1,000 lbs/in^2, or simply 900 lbs/in^2. The *%Span* of the pressure gage would be the accuracy divided by the span.

Accuracy is sometimes specified as *%Reading*. This means that the accuracy of the instrument depends upon the value of the reading being taken. In the example of the pressure gage, suppose the %Reading from the specification sheet was ±0.5%. Further suppose that the gage is now reading 300 lbs/in^2. The accuracy of the gage at this reading is then ±0.005 times 300, or ±1.5 lbs/in^2.

The final method of specifying accuracy to be covered by this text is the *absolute* method. In this case, the absolute accuracy is specified. Again, for the example of the pressure gage, the accuracy might be stated to be ±10 lbs/in^2. Therefore, at a reading of 300 lbs/in^2, the accuracy is ±10 lbs/in^2. At a reading of 800 lbs/in^2, the accuracy is still ±10 lbs/in^2. The accuracy is independent of the value being indicated by the instrument and of the range of values that can be indicated.

EXAMPLE 1-3

Suppose the pressure gage in Example 1-2 has a scale that begins at 3 lbs/in^2. What is the %Span?

$$\%\text{Span} = \frac{\pm 2.5 \text{ lbs/in}^2}{30 - 3 \text{ lbs/in}^2} \times 100\% = \pm 9.26\%$$

What is the absolute accuracy of the gage in Example 1-2?

The absolute accuracy is ±2.5 lbs/in^2, as stated. This amount would need to have been determined by the user or (more commonly) taken from the spec sheet for the instrument.

If the gage in Example 1-2 had no absolute accuracy stated but instead had a %Reading of ±0.4%, what would be its accuracy for a reading of 15 lbs/in^2?

$$\text{Accuracy at 15 lbs/in}^2 = \pm 0.4\% \times 15 \text{ lbs/in}^2$$
$$= \pm 0.06 \text{ lbs/in}^2$$

Precision

Precision refers to the fineness with which the instrument can be read. For example, many calipers have a circular scale with a revolving needle hand, as shown in Figure 1-6. This circular scale can easily be read to the nearest 0.001". Thus, the precision of the instrument is at least 0.001". If it is found that the smallest divisions on the circular scale are far enough apart, one might even be able to determine linear measurement to the nearest 0.0001" or perhaps 0.0002". In these cases, the precision would be 0.0001" or 0.0002", respectively.

This does not mean that the instrument is accurate to 0.001" or 0.0001" or 0.0002". This would have to be determined by using a gaging block whose measurement was known to be accurate to at least 0.0001" (for the 0.001" precision caliper) or 0.00001" (for the 0.0001" precision caliper). A gaging block, or any calibration standard for that matter, should be at least 10 times more accurate that the instrument whose accuracy is being measured. Even though the caliper could read to the nearest 0.001", for example, the reading would not necessarily be correct. It could be in error because the jaws of the caliper are not properly aligned or because there is excess looseness in the moving mechanisms of the instrument. There could be excess play in the gears of the scale. It is often true that an instrument will not be as accurate as the precision with which it can be read.

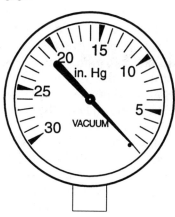

Figure 1-8 A typical vacuum gage.

EXAMPLE 1-4

What is the precision of the vacuum gage shown in Figure 1-8? What is the accuracy?

From the figure of the gage, the authors estimate the precision to be ±0.1 in Hg or simply 0.1 in Hg. This estimate was done by assuming the gage user can estimate the position of the needle between the one-inch-Hg marks to within ±0.1. The accuracy would have to be obtained from the manufacturer's specifications sheet or by calibrating the gage against a known standard.

The precision of a digital display is easier to determine than that of the analog displays discussed so far. The precision is simply equal to the right hand digit in the display. If a digital ammeter has a display capacity in the format of X.XXXX amps, then the precision is 0.0001 amp. Again, this does not mean that the digital ammeter is accurate to 0.0001 amp. It simply means that the ammeter can provide current readings to the nearest 0.0001 amp. It is often true that the accuracy of the digital display instrument is greater than the precision it displays. In this case, the accuracy might be expressed as ±½ count or ±½ digit, meaning that the accuracy is twice as great as the precision of the instrument.

EXAMPLE 1-5

What is the precision of the digital meter in Figure 1-7?
The precision would be ±0.01 unless the not very common case explained in the next paragraph applied to this case.

There are some exceptions to the rule for digital displays. Some displays do not vary the right hand digit by ones. For example, many digital weighing scales will provide measurements only to the nearest half-pound. In other words, starting at zero the scales would be able to output digital displays of 000.0, 000.5, 001.0, 001.5, 002.0, and so on. The precision here would be 000.5 lb. This means that the scales can be read only to the nearest half-pound.

Reproducibility

Reproducibility is the ability of an instrument to produce the same measurement over and over again (at various, but not consistent, times) when the conditions are static. In the case of the caliper, if the jaws were bent it probably would not give accurate readings. This would not affect its reproducibility, however. Whatever inaccuracy the bent jaws entered into the reading would be the same for each new reading taken because the bend in the jaws would give the same error each time. This would not be true of the slack or looseness in the sliding mechanisms or in the circular scale. Slack in a mechanism can cause readings to be off either way from the true value in a random manner. Thus the slack or looseness would cause nonreproducibility in the measurements taken by the instrument.

Nonreproducibility can be caused by factors other than the mechanical slack described previously. For example, frictional drag can cause a pressure gage to be slightly off from the true value. The difference between the true value and the gage reading would depend upon the direction from which the gage needle approached the final reading. If it approached it from below, frictional drag would cause the instrument to read less than the true value. If approached from above, the reading would be greater than the true value. (This is why many pressure gages need to be tapped on their face by a finger before a reading is taken.)

Nonreproducibility that is dependent upon the direction with which the measurement approaches the true value is known as *hysteresis*.

EXAMPLE 1-6

Both actual pressures and measured pressures are given below for two pressure gages that were connected to a pressure calibration device. Do the gages exhibit any nonreproducibility? If so, is the nonreproducibility of the random type or hysteresis type?

True Value lbs/in^2	As Pressure Decreased		As Pressure Increased	
	Gage A	Gage B	Gage A	Gage B
25.0	25.2	25.2	25.3	24.9
20.0	19.7	20.2	20.0	19.9
15.0	15.1	15.2	14.8	14.9
10.0	9.9	10.2	9.8	9.9
5.0	5.1	5.2	5.1	4.9
0.0	0.2	0.2	0.0	0.0

Gage A obviously exhibits nonreproducibility. For example, at one time when the actual pressure is 15.0 lbs/in^2, the gage reads 15.1 lbs/in^2; at another time, it reads 14.8 lbs/in^2. The nonreproducibility does seem to be random, however. When the true pressure is decreasing, the measured values are both too high and too low. If hysteresis were present, the readings would systematically be either too high or too low, but not both. Likewise for the readings taken where the true pressure was increasing.

Gage B on the other hand is consistently 0.2 high on decreased readings and 0.1 low on increased readings. This is typical of a gage with hysteresis.

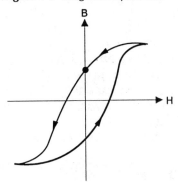

Figure 1-9 Magnetic hysteresis.

Hysteresis can also be caused by fundamental intrinsic physical factors. Magnetic materials are well known to exhibit hysteresis. If the magnetizing force is plotted versus the induced magnetic field in the material, a curve such as that shown in Figure 1-9 is found. The lower portion of the curve represents the result of the magnetizing force being increased from its lowest value, while the upper portion of the curve represents the magnetizing force being decreased from its highest value. The induced field curve on the way down does not follow the same path it took on the way up. This is due to residual magnetism remaining in the material. Some instrumentation devices that incorporate magnetism as part of their functional operation exhibit small amounts of hysteresis.

A few static instrument characteristics that are not as common as the ones discussed so far in the text are resolution, repeatability, and linearity. *Resolution* is the smallest amount of the variable being measured that the instrument can resolve. In practical terms, resolution is the larger of the two characteristics, accuracy and precision. Resolution is most commonly given in absolute form, such as ±1 lb/in^2.

Repeatability is a measure of the closeness of agreement among several consecutive readings taken of a variable. The readings are taken consecutively so that the value of the variable does not have time to change.

Linearity is a measure of the direct proportionality between actual value of the variable being measured and the value of the output of the instrument. The automotive "gas" tank fluid level gage (more commonly known as the "gas gage") is notorious for the problems it causes because of nonlinearity at the bottom end of its scale. The easiest method for determining linearity is to plot the output of the instrument against the actual value of the variable. The plot will be a straight line for a perfectly linear device.

Up until now, the terms *displayed value, indicated value,* and *reading* have been used interchangeably, and rightly so. It should be pointed out, however, that an instrument is not expected to provide a reading—merely an output. This was implied earlier when the definition of the term *instrument* was discussed. For example, when the sensor determines the value of the measured variable in the process control loop, it is not expected to display the value of the variable. It simply needs to provide an output signal to the controller. During all the discussion up to now of the static characteristics of instruments, the term *output* could have just as correctly been used as any of the preceding three terms. It was not used because the beginning student in instrumentation is more comfortable with the idea, for example, of a pressure gage giving a reading than of sending a signal. In future chapters, the term *output* will be used more and more.

REVIEW MATERIALS

Important Terms

instrument	process
process control	manipulated variable
measured variable	controlled variable
element	sensor
transducer	actuator
controller	set point
transmitter	accuracy
percentage full-scale accuracy	span
percent reading	absolute accuracy
precision	reproducibility
hysteresis	resolution
repeatability	linearity

Questions

1. Some instruments simply measure and display while others measure and provide a signal indicating the size of the measurement. (Of course, these latter types can easily have a display of the measurement also.) For the instruments listed below indicate whether they are simply measure-and-display (m & d) devices or are measure-and-signal (m & s) devices.

    Calipers with gage as pictured in Figure 1-6

    Thermocouple

    Voltmeter

    Mercury thermometer

    Photocell

2. What are some physical variables, other than those mentioned in this chapter, that exist in the industrial setting that might need to be measured?

3. Is home central heating an example of process control? If so, what variable is being controlled by the process system?

16 / INTRODUCTION

4. What are some other devices that could serve as actuators?
5. Would an ordinary mercury thermometer be a sensor; a combination sensor and transducer; or a combination sensor, transducer, and transmitter?
6. For a typical home central air conditioning system, what would be the following: A normal set point? The sensor? The measured variable? The manipulated variable?
7. Why might a digital pressure gage be preferable to an analog pressure gage?
8. If a pressure gage had an internal leak, would this affect its accuracy? Its precision? Its reproducibility?
9. Obtain a spec sheet for an instrument. Does the sheet refer to the accuracy of the instrument? Does it refer to any other static characteristics of the instrument?
10. Automotive gas gages are notorious for giving erroneous displays. Which of the static characteristics of instruments mentioned in the chapter most needs to be improved?

Problems

1. For Figure 1-10, identify the process, the measured variable, the manipulated variable, the controlled variable, and the following elements: controller, actuator, and sensor.
2. Draw a schematic (similar to the type in Figure 1-3b) for a home or office central air conditioning system. Label the elements. What would be the measured variable? The manipulated variable? A reasonable value for the set point? Name devices that might serve as elements.

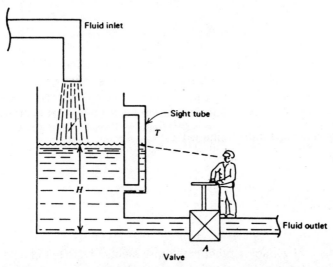

Figure 1-10 Example of manual process control *(From Curtis Johnson, Process Control Instrumentation Technology, 4th ed., © 1993, p. 3. Reprinted by permission of Prentice Hall, Englewood Cliffs, N.J.)*

3. A thermocouple is a heat-sensing device that typically outputs a signal in microvolts, such as 10 microvolts. If the signal transmitted to the controller needed to be 10 volts, what approximate amplification would the transmitter need to supply?
4. An instrument that measures force has a range of 0–250 lbs with a %FS accuracy of ±0.2%. What is the absolute accuracy of the instrument?
5. An instrument that measures force has a range of 0–50 lbs and an accuracy of ±2 lbs. What is its %FS accuracy?
6. A temperature instrument has a range of 300–1,200° F. A reading is taken of 450° F. What is the possible error due to accuracy if the instrument has ±5%FS accuracy? ±5%Span accuracy?
7. A temperature instrument has a range of –40° F to 300° F. Its absolute accuracy is ±3° F. What is its %FS? %Span?
8. A force measuring instrument has a span of 50–250 lbs. Its accuracy is ±1 lb. What is its %FS? %Span? %Reading? Absolute accuracy?
9. What would be the precision of a digital voltmeter with the display format of XX.XX but whose maximum measurable voltage is 19.99 volts? If its accuracy was specified as being ±½ count (or ±½ digit), what would be its accuracy in units of volts?
10. What would be the %FS accuracy of the voltmeter in Problem 9?
11. A pressure instrument was checked for hysteresis, and the data shown in Table 1-1 were obtained. Graph the data so as to see if hysteresis is present and analyze the results.

TABLE 1-1 Data for Problem 11

True Pressure (lbs/in²)	Pressure Increasing Gage Reading (lbs/in²)	Pressure Decreasing Gage Reading (lbs/in²)
30.0	29.9	30.0
25.0	24.7	25.3
20.0	19.7	20.3
15.0	14.6	15.3
10.0	9.7	9.8
5.0	4.8	5.0
0.0	–0.1	0.0

DC Electricity

CHAPTER GOALS

After study of this chapter, you should be able to do the following:

Define the basic terms of electricity such as voltage and current.

State Ohm's Law and use it for all applicable situations.

Identify components of electric circuits such as resistors, capacitors, and inductors.

Understand how voltage can be divided by fixed or variable resistances.

State the various rules for computing total resistance for both parallel circuits and series circuits.

Be able to analyze a Wheatstone bridge, especially when used in conjunction with a strain gage.

State how a strain gage can be used to measure change of position, pressure, load, and the like.

Know the possible effects of electrical meters on the quantities that are to be measured.

Understand that the intrinsic properties of many resistive devices make them suitable for use as sensors/transducers.

Be able to analyze circuits by the application of Kirchhoff's Laws.

One might consider the subjects of electricity, pressure, temperature, level, flow, and humidity to be parallel subjects, especially since they constitute individual chapters in this textbook. Electricity, however, is a tool while the others are physical variables. (Actually, pressure is both a tool and a physical variable.) In the industrial setting, electricity is used to power many of the motors used to drive equipment, as well as to provide lighting and other necessities. In the industrial subfield of instrumentation and process control, electricity is used both to provide power for many instruments and other process elements and to send signals. In particular, it is used to send signals indicating the value of measured variables. It is also used to send signals to actuators indicating the changes needed in manipulated variables. This chapter covers the fundamental aspects of DC electricity needed to understand its use in instrumentation and process control.

BASIC CONSIDERATIONS

Power and information in the form of signals are transported by electricity through the use of the electric fields of mobile electrons in materials. The electrons are caused to move by an electromotive force known as *voltage*. The electrons are able to move through materials with an ease that depends upon the type of material. Metals have a "sea" of electrons that are basically free to move in response to any attractive or repulsive electrical voltage. Materials such as wood, plastic, and glass, on the other hand, have electrons too tightly bound to the atomic nuclei to be free to move at all. These and similar materials are called *insulators*. The measure of the difficulty of electron movement in any given material when voltage is present is known as *resistance*. Materials, such as metals, which have a very low resistance are designated *conductors*. Some devices are made of materials with considerable resistance that still fall in the category of conductors. When their purpose is to provide electrical resistance at points in an electrical circuit, such devices are known as *resistors*. *Superconductors* are materials with zero electrical resistance. Such materials are still in the research stage for most practical purposes.

From the previous definitions, then, *insulators* would be materials of extremely high electrical resistance. One other category is that of the semiconductors. *Semiconductors* are materials whose resistance to electron flow lies between that of insulators and that of conductors.

DIRECT CURRENT

Current is the rate of electron flow through a material. It is measured in units of coulombs per second, C/sec, where a *coulomb* is the unit of electric charge. For example, the charge carried by a single electron is 1.60207 × 10⁻¹⁹ C. The unit, coulombs per second, has been given the shortened name *ampere*, more commonly referred to simply as "amp," which is represented by the symbol A. The unit of electromotive force is the *volt*, symbol v, and the unit of resistance is the *ohm*, symbol Ω.

By an analogy to the flow of water through a pipe, the resistance just mentioned would be equivalent to a restriction in the pipe, such as a partially closed valve. The voltage would be represented by the pressure that is causing the water to flow. Just as voltage is provided by a source such as a generator or battery, pressure is provided by sources such as pumps and gravity (water standing in a tall pipe, for example). The analogy of the wire conductor would be the pipe. The current would be paralleled by the water flow.

In Figure 2-1, the flow is seen to be propelled by the pump. The water has a choice of three different resistive paths through which to flow. Some water will flow through each path, but the path with the least resistance will have the greatest water flow.

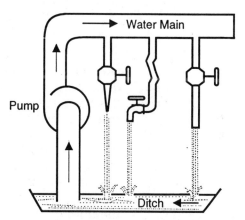

Figure 2-1 There is similarity between a water system and an electric circuit.

In fact, electric circuits can be analyzed in much the same manner as fluid circuits. When fluid flows through a restriction, there is a pressure drop across the restriction. When current passes through a resistance, there is a voltage drop across the resistance.

Although electrical current consists of the movement of electrons from one point to another (and also the repulsive force of electrons acting upon one another), the direction of current flow in most textbooks is taken to be the direction that positive particles would flow if they existed. The in-depth investigation of electricity began about 100 years before the electron was discovered. In these early years, it was assumed that current consisted of the flow of positive particles, and all laws and other discoveries were written and analyzed with this convention in mind. A few textbooks on the subjects of electricity and electronics available today use the convention that current flow is taken to be the direction that electrons would flow, but these textbooks constitute a definite minority. This textbook uses the more common convention of positive particle flow.

Ohm's Law

According to Ohm's Law, a volt impressed across a resistance of 1 ohm will produce a current of 1 ampere:

$$E = IR \tag{2-1}$$

or, after solving for I,

$$I = \frac{E}{R} \tag{2-2}$$

where I stands for current in amps, E for voltage in volts, and R for resistance in ohms. The symbol E is used for voltage in this book instead of the more obvious V symbol because V is reserved for "volume" and "velocity," and also because it is somewhat more common to use E. (The E comes from the term *electromotive force*, which is basically a synonym for voltage.) The unit *volt*, however, is represented with a small v. The unit for resistance, *ohm*, is defined as one volt divided by one amp. If any two of the quantities in Ohm's Law are known, the third quantity can easily be calculated.

EXAMPLE 2-1

If the voltage across a 100 Ω resistor is 10 volts, how much current is passing through the resistor?

$$I = \frac{E}{R} = \frac{10 \text{ v}}{100 \text{ Ω}} = 0.1 \text{ A, or } 100 \text{ milliamps (mA)}$$

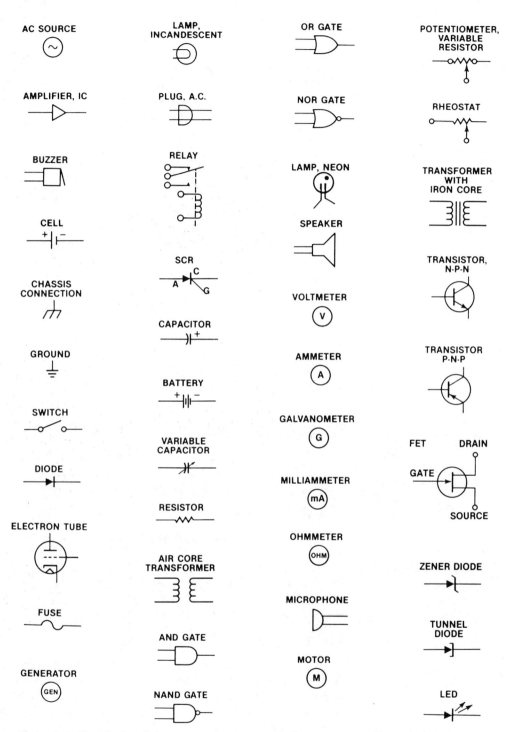

Figure 2-2 Electrical and electronic schematic symbols *(From Rex Miller and Fred W. Culpepper, Jr., Electricity and Electronics, 2nd ed., © 1991, Delmar Publishers, Inc., Albany, N.Y.)*

EXAMPLE 2-2

If the current in a resistor is 0.5 amp and there is a 15 volt change in voltage from one end of the resistor to the other, what is the resistance of the resistor?

$$R = \frac{E}{I} = \frac{15 \text{ v}}{0.5 \text{ A}} = 30 \, \Omega$$

The schematic symbols for a resistor and for a voltage source such as a battery are shown in Figure 2-2, along with many other electrical/electronic symbols. Also shown is the schematic symbol for a voltage cell. A combination of two or more cells is defined as a *battery*.

In Figure 2-3, a voltage source E and a resistor R are shown in an electrical circuit, along with the resulting current I. Note that in Figure 2-3 the current is flowing toward the negative side of the battery.

Figure 2-3 Very simple electrical circuit.

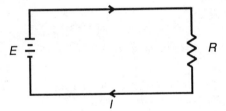

EXAMPLE 2-3

If the current in the circuit shown in Figure 2-3 is 1.2 amps and R is 110 ohms, what is E?

$$E = IR = 1.2 \text{ A} \cdot 110 \, \Omega = 132 \text{ v}$$

Resistor Types and Values

Many different types of resistors are available. Some common types include carbon composition, carbon film, metal film, and ceramic wirewound. These types are shown in Figure 2-4a. The carbon composition resistor is the most commonly used type.

Resistors are available is various sizes, both in regard to physical size and in regard to resistance value. The physical size of a resistor is usually related to the amount of power that the resistor can withstand without overheating and possibly becoming an "open" or a "break" in the circuit. The relationship between the physical size of the carbon composition resistors and the power that each can safely dissipate is shown in Figure 2-4b.

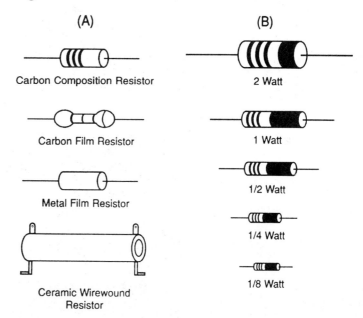

Figure 2-4 **(A)** Types of resistors; **(B)** heat dissipation ability of different sizes of resistors *(Adapted from Jack W. Chaplin,* Instrumentation and Automation for Manufacturing, *© 1992, Delmar Publishers, Inc., Albany, N.Y.)*

Most carbon resistors have colored bands that indicate their resistance value. The color code for these resistors is given in Figure 2-5. The last band in the series of bands (the band closest to the center of the resistor) indicates the accuracy of the color code resistance value. A silver band means that the actual resistance of the device will be within the range +10% to −10% of the color code value. A gold band indicates closer tolerance, ±5%. No fourth band at all indicates a tolerance of ±20%.

EXAMPLE 2-4

A carbon composition resistor has the following band colors starting at one of the ends of the resistor: yellow, blue, orange,

Color	1st Digit	2nd Digit	x By	Tolerance
Black	0	0	1	Gold ± 5%
Brown	1	1	10	Silver ± 10%
Red	2	2	100	No Band ± 20%
Orange	3	3	1,000	
Yellow	4	4	10,000	
Green	5	5	100,000	
Blue	6	6	1,000,000	
Violet	7	7		
Gray	8	8		
White	9	9		

Figure 2-5 Color-code system indicating resistance *(From Jack W. Chaplin,* Instrumentation and Automation for Manufacturing, *© 1992, Delmar Publishers, Inc., Albany, N.Y.)*

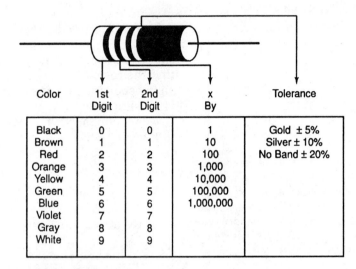

silver. What is its nominal resistance? Within what range is its actual resistance?

The colors yellow, blue, orange, and silver would indicate

$$4 \quad 6 \quad \times \quad 1{,}000 \quad \pm 10\%$$

hence the nominal value of the resistor is 46,000 Ω, and the actual resistance should be within the range 46,000 ± 10% Ω, or 41,400 Ω to 50,600 Ω.

Resistors in Series

When more than one resistor is present in a circuit, the calculation of resistance becomes more complicated, and the laws for resistors in series and resistors in parallel must be used. The total resistance for resistors in series is found by simply adding the individual resistances. In Figure 2-6, where resistors R_1, R_2, and R_3 are in series, the total resistance is

$$R_T = R_1 + R_2 + R_3 \tag{2-3}$$

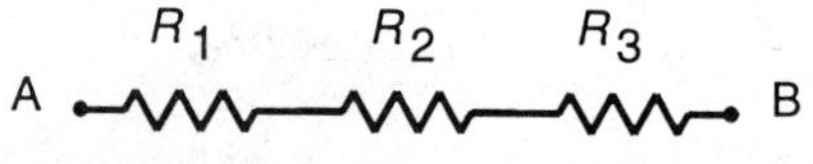

Figure 2-6 Resistors in series.

EXAMPLE 2-5

Calculate the total resistance between points A and B for the resistors in series pictured in Figure 2-7. What is the current through each resistor?

Figure 2-7 Circuit with resistors in series.

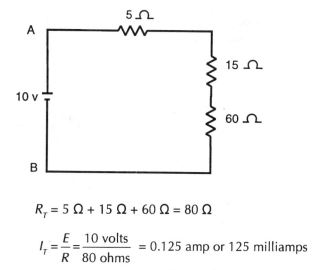

$$R_T = 5\ \Omega + 15\ \Omega + 60\ \Omega = 80\ \Omega$$

$$I_T = \frac{E}{R} = \frac{10\text{ volts}}{80\text{ ohms}} = 0.125 \text{ amp or } 125 \text{ milliamps}$$

Since there is only one path for current in this circuit, the total current must pass through each resistor, and that is 0.125 amp.

Voltage Dividers

Simple dividers. Resistors placed in series are often used as voltage dividers, an arrangement that divides a given voltage into more than one part. The sum of the parts adds up to the original voltage. This idea is explained further in the next example:

EXAMPLE 2-6

Three resistors are in series with a 30 v battery as shown in Figure 2-8. What is the value of the voltage drop across each resistor? (Note: The term *voltage drop* refers to the voltage difference between any two given points. Most commonly these two points are at the beginning and end of a resistance.)

Since the current is the same through each resistor, the source voltage is simply divided into three parts in proportion to each resistor's value. One can skip the effort of calculating the current to find each individual voltage by simply multiplying the source voltage by each resistance in turn, and then dividing by the total resistance.

$$R_{5\Omega} = 30\text{ v} \times \frac{5\ \Omega}{30\ \Omega} = 5\text{ v}$$

Figure 2-8 Series resistors used as a voltage divider.

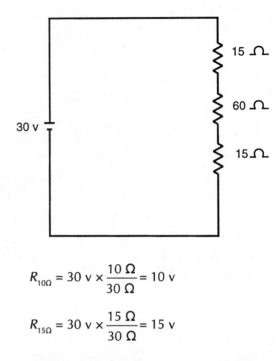

$$R_{10\Omega} = 30 \text{ v} \times \frac{10 \text{ }\Omega}{30 \text{ }\Omega} = 10 \text{ v}$$

$$R_{15\Omega} = 30 \text{ v} \times \frac{15 \text{ }\Omega}{30 \text{ }\Omega} = 15 \text{ v}$$

This method is better understood and appreciated if one calculates the voltage across each resistor using the longer, more tedious route: Calculate the current through each resistor and multiply this current times the value of the resistor.

Note that no point in existence has an absolute voltage. Voltage is always measured with respect to another point. For example, if a battery has a voltage of 12 volts, this means that one terminal is either 12 volts higher or 12 volts lower than the other. However, it is common to denote the earth as "ground" voltage and to give it the value of zero voltage and then to measure other voltages with respect to ground. If the negative or low voltage side of the battery just mentioned were connected to ground, then the positive terminal would be at the voltage level of +12 volts, by convention. Conversely, if the positive terminal were connected to ground, the negative terminal would be at −12 volts. The symbol for a ground connection is shown in Figure 2-2. (Alternatively, ground connections are also very often denoted simply as "0 volts.") Note that oftentimes for troubleshooting purposes a chassis connection (a connection to the metal container of the instrument or circuit) is substituted for a ground connection.

Potentiometers. *Potentiometers* are variable resistance devices that operate by dividing voltages. Basically they use a *slider* or *wiper* to "pick off" voltages at various (usually continuous) points along a fixed resistor upon which the slider rests. An example is shown in Figure 2-9. Here, the resistor shape is almost a full circle and is made up of carbon mounted in a case. The slider is the movable terminal whose movement is controlled by the movable shaft. The symbol for a potentiometer is shown in Figure 2-2.

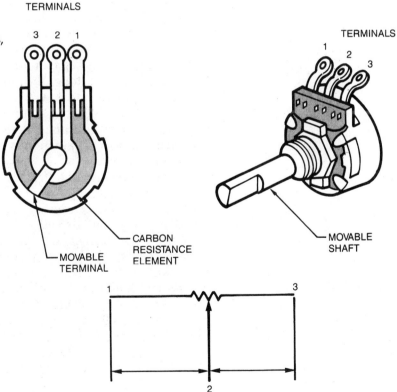

Figure 2-9 Rotational type of potentiometer *(From Earl D. Gates,* Introduction to Electronics, *2nd ed., © 1991, Delmar Publishers, Inc., Albany, N.Y.)*

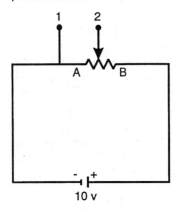

Figure 2-10 Circuit with a potentiometer.

EXAMPLE 2-7

Suppose a total voltage of 10 v is across the potentiometer as shown in Figure 2-10. If the slider is placed in the center position, what voltage will exist between points 1 and 2? If the slider is $1/10$ of the distance from the "A" end of the resistor, what voltage will exist between points 1 and 2?

When the slider is at the midway position,

$$V_{1-2} = 0.5 \times 10 \text{ v} = 5 \text{ v}$$

When the slider is $1/10$ of the distance from the "A" point,

$$V_{1-2} = 0.1 \times 10 \text{ v} = 1 \text{ v}$$

Because the slider "picks off" voltages in proportion to its position along the resistor, potentiometers can detect changes in position of objects, which ultimately allows them to detect changes in physical variables.

The fixed resistor in many potentiometers is in the form of a coil of resistive wire, where the individual turns or coils are just barely separated. The slider then moves along the individual coils, touching only one or two at a time. The voltage output variation will thus be in somewhat discrete steps as compared to the continuous variation provided by the solid carbon resistance. As a result, potentiometers using wire coils are usually not as precise as those using solid carbon resistance.

Potentiometers may be straight in regards to physical shape or they may be circular as in Figure 2-9. In the circular case, the potentiometer can be used to detect not only rotational position change but also other types of position changes that have been transformed to rotational changes through the use of gears, cams, linkages, or some other means. Also, the wire coils can be wound with the use of special wire so that the resistance does not vary linearly with slider position but logarithmetically. These types of potentiometers are more commonly used for audio volume controls, however, than for variable detection. Finally, it should be mentioned that potentiometers are oftentimes referred to simply as *pots*, a shortened form of the name "potentiometer."

Resistors in Parallel

The total resistance for resistors in parallel can be found by following the general formula

$$\frac{1}{R_T} = \frac{1}{R_1} + \frac{1}{R_2} + \frac{1}{R_3} + \ldots + \frac{1}{R_n} \qquad (2\text{-}4)$$

EXAMPLE 2-8

Calculate the resistance between point A and point B as shown in Figure 2-11:

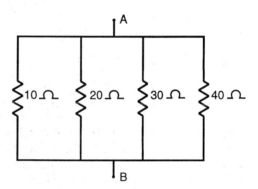

Figure 2-11 Resistors in parallel.

$$\frac{1}{R} = \frac{1}{10\Omega} + \frac{1}{20\Omega} + \frac{1}{30\Omega} + \frac{1}{40\Omega} = 0.208\ \Omega^{-1}$$

$$R = \frac{1}{0.208\ \Omega^{-1}} = 4.8\ \Omega$$

The case of two resistors in parallel occurs so often that it is common to simply memorize the following formula derived from Equation 2-4:

$$R = \frac{R_1 R_2}{R_1 + R_2} \qquad (2\text{-}5)$$

An example of the use of Ohm's Law for a more complex resistance circuit is given next:

EXAMPLE 2-9

What is the value of *I* for the circuit shown in Figure 2-12? What is the value of the current through the 150 Ω resistor?

Figure 2-12 Circuit for Example 2-9.

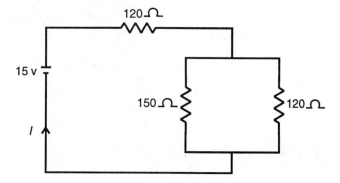

The best way to proceed with a problem of this type is to find the total resistance of the circuit and divide the source voltage by this resistance to get the circuit current. The first step toward finding the total resistance would be to find the equivalent resistance of the two parallel resistors.

$$R_{parallel} = \frac{(150 \text{ }\Omega)(120 \text{ }\Omega)}{150 \text{ }\Omega + 120 \text{ }\Omega} = 66.7 \text{ }\Omega$$

$$R_T = 66.7 \text{ }\Omega + 120 \text{ }\Omega = 186.7 \text{ }\Omega$$

(or 187 Ω rounded off).

$$I = \frac{E}{R} = \frac{15 \text{ v}}{187 \text{ }\Omega} = 0.0802 \text{ A, or } 80.2 \text{ mA}$$

To find the current through the 150 Ω resistor, it will be necessary to know the voltage across it. The easiest way to find this voltage is to find the voltage across the nonparallel 120 Ω resistor and to subtract this voltage from the 15 v source voltage.

$$E_{120\Omega} = IR = (0.0802 \text{ A})(120 \text{ }\Omega) = 9.62 \text{ v}$$

$$E_{150\Omega} = 15 \text{ v} - 9.62 \text{ v} = 5.4 \text{ v}$$

$$I_{150\Omega} = \frac{E}{R} = \frac{5.4 \text{ v}}{150 \text{ }\Omega} = 0.036 \text{ A, or } 36 \text{ mA}$$

Resistance Bridges

Somewhat complex resistive circuits are sometimes built in forms that are commonly called "bridges." The name comes from the location of the measuring meter on the electrical schematic; it "bridges" the sides of the circuit. In other words, one side of the meter is connected to one side of the circuit and the other side of the meter to the other side of the circuit.

Bridges are the electrical circuits that are commonly used with varying-resistance devices to accurately determine any amount of resistance change. The most common bridge is the Wheatstone bridge, shown in Figure 2-13.

When the resistances R_1, R_2, R_3, and R_4 are all equal, the bridge is "balanced," and the voltage at point A is equal to the voltage at point C. The voltmeter in the center would thus read zero. Now, suppose R_1 for some reason were cut in half. The voltage at point A would now become closer to E, the supply voltage, because there is less resistance between point A and point B than before. The voltage difference between points A and C will no longer be zero. It will be a measurable difference, and the voltmeter will indicate this difference. As R_1 continues to increase or decrease, the voltmeter will monitor the changes. Bridge circuits are very sensitive to small changes in resistance. They can also be set up so to compensate automatically for temperature changes that cause unwanted resistive changes in the measuring network. Because of these two features, bridges are used extensively in instrumentation.

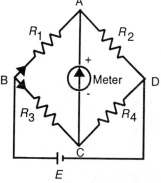

Figure 2-13 Wheatstone bridge.

EXAMPLE 2-10

Assume that all four resistors in the Wheatstone bridge shown in Figure 2-13 have the value of 50 ohms except R_1, which has the value 25 ohms. E is 12 volts. What will the voltmeter read?

The voltage at point C will be 6 volts because $R_3 = R_4$ and the total voltage drop across the two resistors is 12 volts. The voltage at point A will be given by

$$E_{BA} = E_T - \frac{(R_1 \times E_T)}{R_1 + R_2} = 12 - \frac{(25 \times 12)}{25 + 50} = 12 - 4 = 8 \text{ v}$$

Assuming that the positive side of the voltmeter is connected to point A, then

$$E_{A-C} = 8 \text{ v} - 6 \text{ v} = 2 \text{ v}$$

EXAMPLE 2-11

Assume that R_1, R_2, and R_3 in the Wheatstone bridge of Figure 2-13 are 250 Ω resistors and that E is 10 v. What would be the value of R_4 if the voltmeter indicates −0.5 v?

The voltage at point A would be 5 v since R_1 and R_2 split the 10 v source voltage equally. Thus, the voltage at point C must be 5.5 v, since it was given that $E_{A-C} = -0.5$ v. To solve for R_4, a voltage divider equation is set up:

$$E_C = \frac{R_4}{R_4 + R_3} \times E = \frac{R_4}{R_4 + 250\,\Omega} \times 10\text{ v} = 5.5\text{ v}$$

Solving for R_4, $R_4 = 306$ Ω, rounded to three significant figures.

Resistive Sensors/Transducers

Wheatstone bridges are used in instrumentation because they are extremely sensitive to small changes in resistance. Thus, they are very often used with resistive devices that are used for sensors or transducers. One such device is the *strain gage*, an example of which is shown in Figure 2-14.

Figure 2-14 An enlarged view of a strain gage *(From Jack W. Chaplin,* Instrumentation and Automation for Manufacturing, *© 1992, Delmar Publishers, Inc., Albany, N.Y.)*

A strain gage is a resistance path made up of copper or nickel particles cemented to the surface of a pad. When the pad is bent or strained, the total resistance of the path changes. If the pad is bent in the direction of the resistive surface so that the surface is now concave, the particles are compressed and the resistance decreases. If the pad is bent so that the surface is convex, the particles are slightly separated so that the resistance increases. The total effect is that the amount of resistance change is a measure of the amount of bending undergone by the strain gage.

The resistance path on a strain gage mat usually consists of "long" (in a relative sense) lines running the length of the strain gage, with sharp turns at each end. This makes the strain gage sensitive to bending only along one axis. Figure 2-15 gives a clearer picture of this than does Figure 2-14.

Strain gages are used in many applications, both separately and built in as an integral part of some device. An example of this latter case are load cells, the principal elements of which are strain gages. (Load cells are discussed further in Chapter 11.)

Other examples of a sensor/transducer that uses stain gages are some types of differential pressure transducers. Such a device has an internal diaphragm with at least one strain gage cemented to the diaphragm. As the two pressures combat one another, they bend the diaphragm. As the diaphragm is bent, the resistance of the strain gage varies. This variation in resistance can be distinguished easily by a Wheatstone bridge circuit which is oftentimes built into the transducer.

Figure 2-16 shows a cross-sectional sketch of such a transducer, with the strain gage cemented to the high-pressure

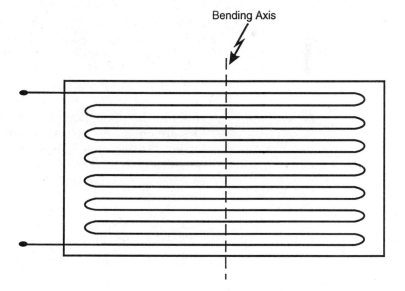

Figure 2-15 A more simplified view of the construction of a strain gage.

Figure 2-16 Electrical and physical placement of a strain gage in a differential pressure transducer.

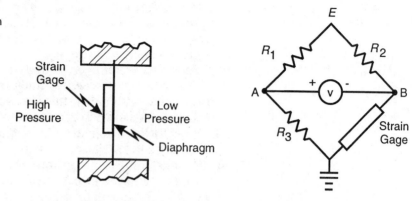

side of the diaphragm. The strain gage is also shown as electrically being one of the quadrants of a Wheatstone bridge. When the high pressure increases with respect to the low pressure, the strain gage is bent in a concave manner, and its resistance decreases. In the Wheatstone bridge, when the resistance of the strain gage decreases, then point B becomes closer to ground (in a voltage sense, that is), thus the voltage at point B lessens. The voltage at point A remains the same because R_1 and R_3 are fixed (as is R_2, of course). Since B drops in voltage value and since the negative lead of the voltmeter is attached to B, the voltmeter will read more volts (assuming point A was already at a higher value than B). When the high pressure decreases with respect to the low pressure, the opposite analysis holds and the voltmeter will read a smaller value.

Note that the situation indicated in Figure 2-16 works just as well with the strain gage in any of the four quadrant locations. If the gage is placed in the upper left quadrant instead of the lower right, the voltmeter reading will vary as indicated in the preceding paragraph. In fact, if all three resistors are equal, the voltmeter output will be exactly the same for both situations. If the strain gage is placed in either the upper right or lower left quadrants, the voltmeter's change of reading will be the exact opposite of the two positions just discussed.

Electrical Meters

Most analog (nondigital) voltmeters and ammeters use some form of electromechanical movement to measure the electrical variable in question. Construction of one such movement type, known as the D'Arsonval meter movement, is shown in Figures 2-17 and 2-18. In these figures, current from the circuit to which the meter is attached flows through the movable coil, producing an electromagnetic field. This field is repulsed by the existing permanent magnetic field, a field that is present because of the permanent magnet built into the meter. The repulsion causes the coil to turn against the force of the spring as its electromagnetic field tries to align with the permanent field. The strength of the electromagnetic field created by the coil is proportional to the amount of current flowing through the coil. Thus, the distance the movable coil can rotate against the spring's force depends upon the current. This means that the rotation of the coil as measured by the pointer attached to it is a function of the amount of current sensed or used by the meter.

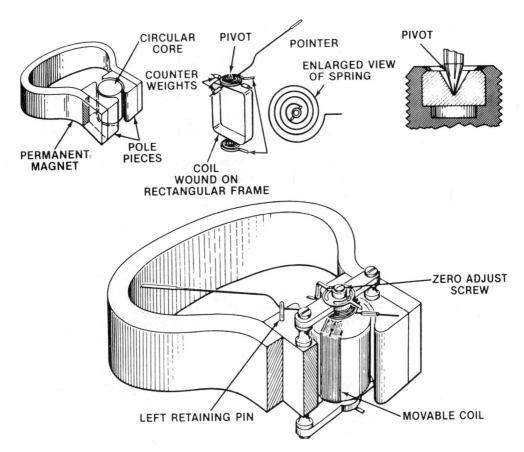

Figure 2-17 D'Arsonval meter movement *(From Rex Miller and Fred W. Culpepper, Jr.,* Electricity and Electronics, *2nd ed., © 1991, Delmar Publishers, Inc., Albany, N.Y.)*

Figure 2-18 Expanded diagram of main portion of D'Arsonval meter movement *(From Rex Miller and Fred W. Culpepper, Jr., Electricity and Electronics, 2nd ed., © 1991, Delmar Publishers, Inc., Albany, N.Y.)*

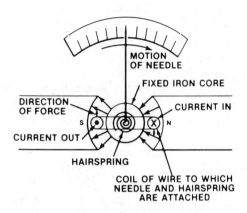

The principal difference between the voltmeter "meter" and the ammeter "meter" is whether internal resistance will be placed in series with the coil or in parallel with the coil. An ammeter must always be placed *in* the actual circuit to perform its measurement. To prevent the ammeter coil from having to carry the entire amount of circuit current during a current measurement, a parallel resistance, known as a *shunt*, is placed around the coil. The ammeter coil thus only reads a small fraction of the actual current, but since this fraction remains constant, the meter can be calibrated to compensate for this.

The voltmeter, on the other hand, is hooked across a circuit voltage difference and only "samples" a small portion of the circuit current because large internal resistance is always in series with the voltmeter coil. The voltmeter is essentially *outside* the circuit. Since the meter's internal resistance is fixed (for any given meter voltage setting), the amount of current passing through the voltmeter coil depends on the amount of voltage difference across which the meter is connected. In a multimeter, which is a combination voltmeter/ammeter, the resistance in series with, or parallel to, the coil is controlled by the function switch on the front of the meter.

In the bridge circuits described in the preceding section, the voltmeter was assumed to have no effect upon the circuit it was measuring. Since voltmeters sample or use a small portion of the circuit current, this is not actually true for any type of circuit. If the voltmeter in Figure 2-19a has an internal resistance less than infinity (which, of course, it must), then current can flow through it (and needs to if the voltmeter needle is to move). The actual circuit as seen by the current is shown in Figure 2-19b.

Figure 2-19 (A) Circuit showing a voltmeter connected across a resistor; (B) circuit showing the voltmeter as seen by the circuit current.

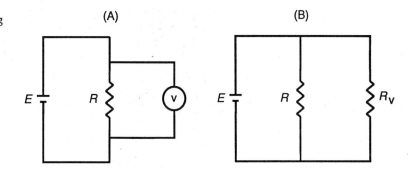

Figure 2-20 Wheatstone bridge as seen by the current.

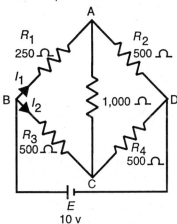

Similarly, the actual circuit seen by the current for the bridge in Figure 2-13 is shown in Figure 2-20. Suppose that the fixed resistors R_2, R_3, and R_4 are 500 ohms, that $E = 10$ volts, and that the voltmeter has an internal resistance of 1,000 ohms. Further suppose that the variable resistor, R_1, is currently at 250 ohms. The voltage at point A would be a fraction, [500/(250 + 500)], of 10 volts or 6.667 volts if there were no pathway connecting one side of the bridge to the other. But there is a pathway; and the resulting path that current can take as it flows from point B to point D along the "A" side of the bridge is shown in Figure 2-21. The actual voltage at point A is [375/(250 + 375)] × 10 volts or 6.000 volts—quite a change from the goal of 6.667 volts. This would lead to considerable error, especially for a device like a bridge circuit which

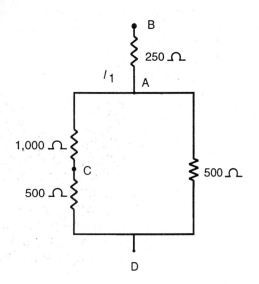

Figure 2-21 Path of the current for the A side of the Wheatstone bridge in Figure 2-18.

is built to measure small changes in resistance. The way to prevent this error from occurring is to select voltmeters that have large internal resistances. Actually, most voltmeters sold today do have large input resistances, digital voltmeters usually having the larger resistances. (An input resistance of ten Megohms is a typical value for a digital voltmeter.) Care should be taken when selecting a voltmeter for any given circuit.

EXAMPLE 2-12

For the example discussed in the immediately preceding paragraph, what voltage would be measured at point A if the voltmeter had an input resistance of 10 Megohms?

Using Figure 2-20 with a 10,000,000 Ω resistor replacing the 1,000 Ω resistor, the resistance of the parallel portion of the network would be as follows, using Equation 2-5:

$$R = \frac{(1{,}000{,}500 \ \Omega)(500 \ \Omega)}{1{,}000{,}500 \ \Omega + 500 \ \Omega} = 499.75 \ \Omega$$

The voltage at point A would then be

$$E_A = \frac{499.75}{499.75 + 250} \times 10 \text{ v} = 6.666 \text{ v}$$

To four significant figures, the voltmeter has introduced no significant error into the original circuit.

The same analysis also applies to ammeters, except that in this case one seeks low resistance. Ammeters are placed in *series* in a circuit in order to measure current while voltmeters are placed in *parallel* in order to measure voltage. Since an ammeter is actually in the circuit, then the less resistance it has, the less the circuit feels the effect of the ammeter's presence. In selecting ammeters, one should choose those with the least internal resistance, balancing this resistance against cost.

Digital meters use electronic circuits instead of the coil and magnet approach. Otherwise, most of the same characteristics apply in regard to the difference between a voltmeter and an ammeter and in regard to the effect of the meter upon the variable it is intended to measure. Digital meters, as a rule, do have less effect upon these variables because of advanced design. Still, the effect of the internal resistance of the meter must always be at least considered, even though it usually has a negligible impact.

If one is using a multimeter, a voltmeter/ammeter combination mentioned earlier, there is usually an *ohmmeter* function included among the meter's selectable functions. When in ohmmeter mode, the multimeter measures resistance. To measure resistance, the meter supplies its own current from an internal battery. An ohmmeter is typically connected to each end of a resistive device and places about 1.5 v across this device. It then measures the amount of current that can flow through this component. Since the voltage across the component is known to be 1.5 v, the amount of current that can pass through it is inversely proportional to the resistance, according to Ohm's Law. For analog ohmmeters, the resistance can be read directly from the calibrated face plate on the meter.

Unlike voltmeter and ammeter functions, ohmmeters usually conduct electrical measurements on disconnected, "nonlive" circuits. Ohmmeters measure the resistance of circuit components, usually with the components removed from the circuit. They are also used occasionally to measure the resistance of most of a whole circuit.

Resistivity

Resistive circuit components have physical characteristics such as length and cross-sectional area. The relationship between the physical characteristics of a resistance element and its overall resistance is most commonly expressed by Equation 2-6:

$$R = \frac{\rho l}{A} \tag{2-6}$$

where l is the length of a device, A is the cross-sectional area, and ρ is a constant known as the *resistivity*. Basically, the equation says that the resistance of a device is directly proportional to the length of the device and inversely proportional to the cross-sectional area. These conditions, especially the condition regarding the effect of length on resistance, make resistive devices useful for monitoring variables such as pressure and position/displacement. In many cases, the constant of proportionality, the resistivity, is a well-defined function of temperature. This characteristic makes special resistive devices very useful for measuring temperature. Figure 2-22

Figure 2-22 Resistance of some various metals.

OHMS RESISTANCE PER MIL-FOOT
(AT 70° F) "K"

Aluminum	17
Brass	42
Bronze (Cu, Sn)	108
Chromel (Ni, Cr)	420-660
Copper	10.4
German silver	200
Gold	14.6
Graphite	4,800
Iron, pure	59
Lead	132
Mercury	575
Nickel	42
Nichrome (Ni, Cr)	550-660
Platinum	60
Silver	9.6
Steels	72-500
Tungsten (wolfram)	33

Figure 2-23 Temperature dependence of resistance.

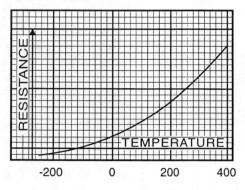

shows how resistivity varies from metal to metal, and Figure 2-23 gives an example of how the resistivity for one metal varies with regard to temperature. Some of these cases will be discussed in later chapters.

Kirchhoff's Laws

There are two laws, known as Kirchhoff's Laws, that are extremely useful for analyzing electrical circuits. The first one is *Kirchhoff's Voltage Law*. This law says that for any closed path in a circuit, the algebraic sum of the voltages is zero. (Alternatively, this law is more simply stated, "The total voltage drops along any series path must equal the source voltage.") Consider Figure 2-24. Because of the polarity of the voltage source, the current will flow in the direction indicated (since positive particles would want to flow away from the positive side of the source and toward the negative side). The voltage dropped across the resistor will have the polarity indicated and will have the same magnitude as the source. Why? Because there is no other resistance available in the circuit across which some portion of the source voltage might be dropped. In other words, there is no voltage divider present—only one single resistor; and thus all the source voltage is going to be dropped across this resistor. The voltage dropped across the resistor must have the polarity indicated; otherwise, there would be, in effect, two voltages in series. The resistor would be acting as a voltage source instead of a voltage user. Since this is not the case, the polarity of the voltage across the resistor must be as shown. Hence, the algebraic sum of the voltages in the loop is zero as one considers them in order of their appearance, starting at any beginning point.

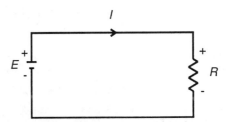

Figure 2-24 A simple example of Kirchhoff's Voltage Law.

To make analysis of electrical circuits easier, one should remember that the end of a resistance the current enters will be positive with respect to the other end of the resistance.

Next, consider a circuit with two voltage sources as shown in Figure 2-25. Which way will the current flow? In more complex circuits such as this, even if the actual voltages are known, the direction of the current cannot always be guessed correctly before the problem is solved. To solve the problem then, it is necessary to assume that the current is flowing in a given direction and mark it as such. Once the problem is solved, if any current values end up being negative, this indicates that the initial assumption in regard to direction was wrong. The current is actually flowing in the opposite direction.

Before dealing with a circuit such as that shown in Figure 2-25, it would be preferable to introduce *Kirchhoff's Current*

Law. This law says that at any node (junction of three or more wires) in the circuit, the sum of the currents entering the node equals the sum of the currents leaving it. More simply stated, the total current coming to any point in a circuit must equal the current leaving that point. This law can be illustrated by using both it and Kirchhoff's Voltage Law to solve for the unknown in Figure 2-25.

Figure 2-25 Circuit for use with Example 2-13.

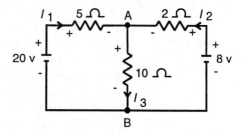

EXAMPLE 2-13

Find the three currents indicated in Figure 2-25.
Applying Kirchhoff's Current Law to node "a,"

$$I_3 = I_1 + I_2 \tag{1}$$

Now, applying Kirchhoff's Voltage Law to both the left loop and the outside loop,

$$20 = I_1(5) + I_3(10) \tag{2}$$

$$20 = I_1(5) - I_2(2) + 8 \tag{3}$$

Note that in Equation 3 the 20 volts is the main voltage and that the equation has been set up to state that all the remaining voltages around the outside loop must add up to 20 volts. If the voltages are opposing the 20 volts in the circuit, then they are entered as positive values on the right hand side of Equation 3. If they are in series with the 20 volts (i.e., nonopposing), then they are entered as negative. An alternative way to have set up Equation 3 would have been to count both voltages as source voltages and put them both on the left hand side but opposing one another:

$$20 - 8 = 5I_1 - 2I_2$$

There are now three equations and three unknowns. Probably the easiest path to solution is to take Equation 1 and substitute it in Equation 2, giving

$$20 = 15I_1 + 10I_2 \qquad (4)$$

Now, add Equation 4 and Equation 3, after combining terms in Equation 3 and multiplying it by a -3,

$$20 = 15I_1 + 10I_2 \qquad (4)$$
$$-36 = -15I_1 + 6I_2 \qquad (3)$$

$$-16 = 16I_2$$

$$I_2 = -1 \text{ amp}$$

Solving for I_1 and I_3 gives

$$I_1 = 2 \text{ amps}$$

$$I_3 = 1 \text{ amp}$$

Note that the negative value for I_2 indicates that this current is actually flowing in the opposite direction from that marked on the diagram.

REVIEW MATERIALS

Important Terms

voltage
resistance
conductor
battery
voltage drop
voltage divider
Wheatstone bridge
strain gage
ammeter
multimeter
resistivity
Kirchhoff's Current Law

current
Ohm's Law
insulator
cell
electromotive force
potentiometer
internal resistance
voltmeter
ohmmeter
digital meter
Kirchhoff's Voltage Law

Questions

1. For any resistor in a circuit, such as Figure 2-3 for example, there will always be a voltage difference across the resistor. Which end of the resistor will have the higher voltage—the end toward the positive side of the voltage source or the end toward the negative side?
2. What are some other materials that might possibly be used to make resistors besides carbon?
3. What voltage sources are available other than the ones mentioned in the chapter: the cell, the battery, and the generator? Which would probably be the most common in the real world of instrumentation and process control?
4. If a high resistance and a low resistance are in series, which will have the larger voltage drop across it? If they are in parallel?
5. If a high resistance and a low resistance are in parallel, what will be the relative value of the total resistance: higher than the high resistance, lower than the low resistance, or between the two resistances? Explain why.
6. If two identical resistors, each of value R, are in parallel, what will be the total resistance?
7. A linear potentiometer can certainly be used to measure physical movement in a straight line. Can a rotational potentiometer be used to do this?
8. Could an ammeter be used in place of a voltmeter in a Wheatstone bridge?
9. Why would a logarithmic potentiometer be used in audio volume situations?
10. How might a logarithmic potentiometer be constructed?
11. Can a negative current exist?
12. What type of meter would one use to measure the resistance value of a resistor?
13. Which electrical meters are self-powered from an internal source such as a battery, and which meters draw their power from the circuit to which they are attached?
14. Which is better, a voltmeter with an internal resistance of 10 ohms or a voltmeter with an internal resistance of 10,000 ohms? Which internal resistance would be better for the case of an ammeter?
15. What would be the steps in measuring the resistance of the resistor in Figure 2-3?
16. If a resistor is made from a piece of wire and the length of the wire is doubled, what happens to the resistance? What happens to the resistance if the cross-sectional area of the wire is doubled?
17. Which would present the most resistance to the flow of current—a cube of copper six feet in length on all sides or a rod of copper six feet long and ¼ inch in diameter?
18. What are the five most conductive metals? Which metal would have the highest resistivity (ρ)?
19. Is a material's resistivity (ρ) a constant under all conditions?
20. Placing the strain gage in Figure 2-16 in what other quadrant of the Wheatstone bridge would present exactly the same results as being in the lower right quadrant as indicated?

21. Suppose a second strain gage were cemented on the opposite side of the diaphragm from the one indicated in Figure 2-16. Where should this strain gage be installed in the Wheatstone bridge to double the output of the initial set-up?
22. Suppose two strain gages were glued to each side of the diaphragm in Figure 2-16 (for a total of four strain gages mounted on the diaphragm). Where would each strain gage on the diaphragm be placed schematically in the Wheatstone bridge for maximum effect whenever the differential pressured varied?

Problems

1. What is the range of possible resistance values for resistors with the following color code bands: (a) yellow–violet–brown–silver? (b) orange–orange–black–no fourth band? (c) white–red–orange–gold?
2. What is the effective resistance between points A and B for (a) Figure 2-26? (b) Figure 2-27?

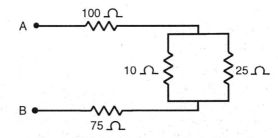

Figure 2-26 Figure for use with Problem 2.

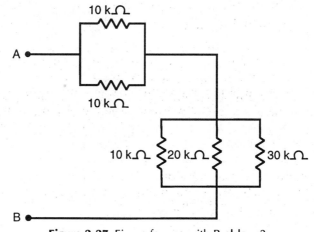

Figure 2-27 Figure for use with Problem 2.

3. Use Ohm's Law to calculate the current I for the circuit shown in (a) Figure 2-28 and (b) Figure 2-29.

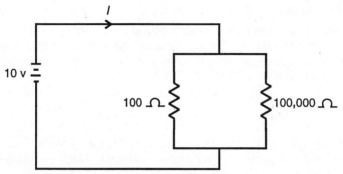

Figure 2-28 Figure for use with Problem 3.

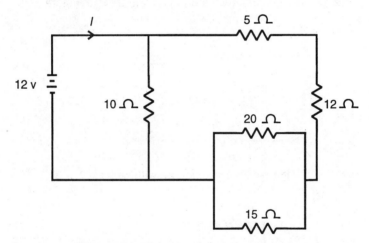

Figure 2-29 Figure for use with Problem 3.

4. Use Ohm's Law to calculate the current flowing through the 100,000 Ω resistor in Figure 2-28.
5. Assuming that the current I has been calculated for the circuit in Figure 2-29 (from Problem 3b), calculate the following: (a) the current through the 10 Ω resistor, (b) the current through the 5 Ω resistor, (c) the voltage drop across the 5 Ω resistor, (d) the voltage drop across the 12 Ω resistor, and (e) the voltage drop across the parallel set of resistors.
6. Calculate the voltage across each individual resistor in Figure 2-30.
7. If a voltmeter were connected to points A and B in Figure 2-31, what would be its reading? Between points C and D? Between points B and D?
8. Suppose point D of the circuit in Figure 2-31 is connected to ground. What would be the absolute voltage at point B? $I_1 = 1$ A.

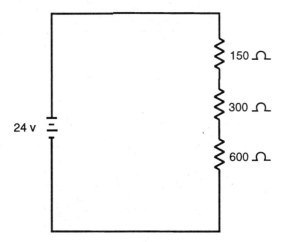

Figure 2-30 Figure for use with Problem 6.

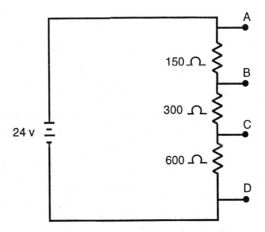

Figure 2-31 Figure for use with Problem 7.

9. Figure 2-32 shows a linear potentiometer connected to a 12 v power supply. (a) If the slider is at the halfway point on the resistance, what is the voltage across points A and B? (b) If the slider is $1/3$ of the way down from the top of the resistance? (c) If the slider is 20% of the way up from the bottom of the resistance?

10. What would be the reading of the voltmeter in the bridge of Figure 2-13 for the following situations (assume the voltmeter has infinite internal resistance): (a) $R_1 = R_2 = 100\ \Omega$, $R_3 = R_4 = 50\ \Omega$, and $E = 10$ v? (b) The same values as part (a) except $E = 24$ v? (c) $R_1 = R_2 = 50\ \Omega$, $R_3 = R_4 = 100\ \Omega$, and $E = 10$ v? (d) $R_1 = R_4 = 100\ \Omega$, $R_2 = R_3 = 50\ \Omega$, $E = 10$ v? (e) $R_1 = R_3 = R_4 = 100\ \Omega$, $R_2 = 90\ \Omega$, $E = 10$ v? (f) $R_1 = R_2 = R_4 = 100\ \Omega$, $R_3 = 90\ \Omega$, $E = 10$ v?

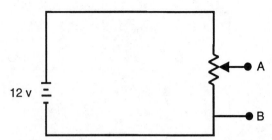

Figure 2-32 Figure for use with Problem 9.

11. Suppose the three resistors in the Wheatstone bridge circuit shown in Figure 2-16 all have the value of 50 ohms. The strain gage has a nominal value of 50 ohms also, when the high pressure is the same as the low pressure. Suppose the high pressure increases and the strain gage's resistance changes by 5 ohms. (Which way would it change? Increase or decrease?) What would have been the voltmeter's initial reading and what would be its new reading? The supply voltage, E, is 12 volts.

12. Rework Problem 11 with the three resistors being 1,000 ohm resistors but all else the same. (The strain gage is still a nominal 50 ohms and changes by 5 ohms.)

13. Suppose it is necessary to measure the voltage across a 10 kΩ resistance. What minimum internal resistance would the voltmeter need to have to keep measurement error to less than ±1.0%? (Measurement error in this case refers to any error caused by the meter affecting the circuit to which it is connected.)

14. The temperature dependence of the resistivity of a metal can often be satisfactorily expressed by $\rho = \rho_0(1 + \alpha T)$, where ρ_0 is the resistivity at 0° C, α is a constant known as the temperature coefficient of resistivity, and T is the Celsius temperature. Plot a graph of the temperature dependence of the resistivity of copper in units of ρ_0. For copper, $\alpha = 0.00393°$ C^{-1}.

15. Find I_1, I_2, and I_3 for the circuit shown in Figure 2-33.

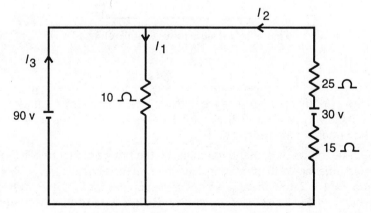

Figure 2-33 Figure for use with Problem 15.

3

AC Electricity

CHAPTER GOALS

After study of this chapter you should be able to do the following:

Discuss the difference between AC circuits and components and DC circuits and components.

Describe how capacitors and inductors can be used to measure physical variables such as movement.

Calculate effective total capacitance for capacitors in series and in parallel.

Discuss the resistance to AC current flow presented by both capacitors and inductors.

Design a very simple circuit to provide a given time delay.

Understand the difference between inductance and reluctance.

Explain the concept of phase.

Calculate electrical power dissipation or consumption for simple situations.

Use your knowledge of electricity to understand the workings of many instruments and control systems.

So far the discussion in this textbook has been entirely about direct current, commonly known as DC, current that flows in one direction only. Most of the concepts presented so far also apply to *alternating current* (*AC*), current that flows in one direction during half of a very short interval of time known as a *period* and the opposite direction during the other half of the period. (A *period* is thus the time it takes a variable to complete one full cycle of variation.) The current is caused to alternate its direction of flow because the voltage source alternates its polarity. The number of times per second the current or voltage completes a cycle is known as the *frequency*. A common example of this type of voltage source is a generator. A simple example of an AC circuit is shown in Figure 3-1.

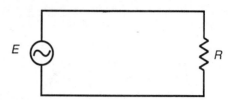

Figure 3-1 Simple example of an AC circuit.

BASIC CONSIDERATIONS

The unit of frequency for an AC source is cycles/second, better known as *hertz (Hz)*. For example, the usual line voltage frequency furnished by the electrical power utility companies in the United States is 60 Hz. This means that the voltage rises and falls through 60 complete cycles every second. Also, if the utility company line voltage is plotted against time, it usually appears as a sine wave. This is discussed further later in this chapter.

To this point in the book I and E have represented DC current and voltage. Since the current and voltage vary with time in AC circuits, the symbols $I(t)$ and $E(t)$ shall represent the instantaneous current and voltage, that is, the value of the current and voltage at a given instant. For practical purposes, however, the analysis of AC circuits is nearly always carried out using effective values of the instantaneous current and voltage. This is done because the effective current and voltage can be used in Ohm's Law. Throughout this textbook when an AC situation is under discussion, the terms I and E represent the effective values of the current and voltage. These so-called effective values are very close to being the average value of the variable over the positive portion of its cycle. The formal mathematical definition of effective values is delayed until the section later in this chapter on phase change.

The resistance to the flow of AC current is shown in Figure 3-1 as an ordinary resistor. Resistors provide opposition to AC current in exactly the same manner and amount as they do to DC current. Other devices also provide opposition to AC current, however, that do not need to be taken into account for

DC circuits. These devices provide opposition to AC current that is dependent upon the frequency of the current. This opposition is given the more general name of *impedance*. Henceforth, the term *resistance* shall mean opposition to current that is not dependent on frequency. Some common and very useful electrical devices provide impedance.

ALTERNATING CURRENT

Capacitance

Capacitance is the measure of a device's capacity to store an electrical charge, where the device is a sandwich of conductor–insulator–conductor materials. The device is known as a *capacitor*. The simplest circuit with a capacitor would be the DC circuit shown in Figure 3-2. There is no current because the capacitor is "open," that is, there is no current path through it. When the DC voltage source is first connected to the capacitor, the positive side of the voltage source attracts electrons from the metallic plate to which it is directly connected. This leaves the plate with a net positive charge since it was neutral beforehand. This positively charged plate then attracts electrons from the negative side of the voltage source. This continues until there are finally enough electrons on the negative side of the capacitor to match the lack of electrons on the positive side of the capacitor. The final voltage across the capacitor ends up being equal to the voltage of the voltage source. All this occurs very rapidly once the voltage source is connected to the capacitor, the time varying according to the resistance in the circuit between the capacitor and the source.

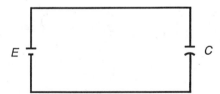

Figure 3-2 Simple capacitive DC circuit.

Although capacitors do not conduct DC current (unless they "leak" or "short"), they do conduct AC current. As the source voltage swings from positive to negative, the steps just outlined occur again—this time in reverse. In fact, because of the continual increase and decrease of the source voltage, the current into and out of the capacitor is continually increasing and decreasing. Capacitors present an infinite resistance to DC current but much less resistance to AC current. The amount of resistance, actually known as impedance, that they do provide in an AC circuit is discussed shortly.

EXAMPLE 3-1

What will probably be the result if a 100 v DC source is used for *E* in the circuit shown in Figure 3-2? What might be the result if a 100 v AC source is used instead?

Probably nothing will happen when the 100 v DC source is connected across the capacitor because no current can flow through the capacitor, and the capacitor will simply have a voltage of 100 v across it. There is some possibility that the 100 v could short across the insulating dielectric between the capacitor plates, however, depending upon the insulating strength of the dielectric. (Capacitors are always rated in regard to the amount of DC voltage they can withstand without "shorting," as well as for their electrical capacitance.)

In the case of the 100 v AC, the capacitor will conduct the AC current. The source might self-destruct in some way. For example, a generator might burn an internal coil, a house wall socket might spark and throw a breaker at the central breaker-box. Whether any of this happens depends on how much current the capacitor conducts, and that depends on the frequency of the AC source. This is further discussed in the next few paragraphs.

The ratio of the amount of charge that can be stored on one of the plates of the capacitor to the amount of voltage across the capacitor is known as the *capacitance*. Stated differently, the amount of charge that can be stored in a capacitor per unit voltage across the capacitor is measured by its *capacitance*—its capacity to store a charge. It has been found that capacitance is directly proportional to the area of the plates and inversely proportional to the distance between them, as given in Equation 3-1:

$$C = \frac{\epsilon A}{d} \qquad (3\text{-}1)$$

The constant of proportionality, ϵ, known as the permittivity, is a function of the type of material separating the plates. Capacitance is measured in *farads* (F). One farad is the amount of capacitance needed to store one coulomb of charge on its plates with a difference of one volt between the plates.

This dependence of capacitance upon the physical properties of area and distance makes this electrical quantity a useful tool for determining such variables as position, pressure, and level. For example, one plate might be stationary and the other plate attached to the object of interest. By using an electrical AC circuit especially built to measure capacitance (usually available simply as a *capacitance meter*), one could determine the relative position of the movable plate by monitoring how much the capacitance of the dual plate setup was changing because of increases or decreases in the distance between the plates.

EXAMPLE 3-2

The plates of a capacitor are 0.05 inch apart and 2 in² in area each. The plates are in vacuum. Given that the permittivity of vacuum is 8.85 × 10⁻¹² F/m, calculate the capacitance.

$$C = \frac{\epsilon A}{d} = \frac{8.85 \times 10^{-12} \ F/m \times 0.0254 \ m/in \times 2.0 \ in^2}{0.050 \ in}$$

$$= 9.0 \ \mu\mu F \text{ or picofarad}$$

What would happen to the capacitance if one of the plates was moved 0.05 inch further away from the other plate?

The capacitance would decrease by half, as indicated by Equation 3-1.

The obstruction, X_c, that a capacitor presents to an AC current (assuming no other devices such as resistors and the like are present in the circuit) is dependent upon its capacitance, C, and upon the frequency, f, of the AC current, as shown in Equation 3-2.

$$X_c = \frac{1}{2\pi f C} \tag{3-2}$$

This AC current obstruction is given the specific name *capacitive reactance*. Capacitive reactance affects AC current in the same algebraic fashion that resistance affects AC or DC current only if no other types of current-impeding devices are present in the circuit. Mathematically, for a simple capacitive circuit, the relationship between AC current, the AC voltage across the capacitor, and the reactance of the capacitor is still in Ohm's Law form, as shown by Equation 3-3:

$$E = I X_C \qquad (3\text{-}3)$$

EXAMPLE 3-3

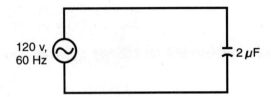

Figure 3-3 Circuit for use in Example 3-3.

Given Figure 3-3, calculate the amount of AC current flowing in the circuit.

$$X_C = \frac{1}{2\pi f C} = \frac{1}{2\pi \cdot 60 \text{ cycles/sec} \cdot 2.0 \times 10^{-6} \text{ F}}$$

$$= 1.3 \times 10^3 \ \Omega, \text{ or } 1.3 \text{ k}\Omega$$

$$I = \frac{E}{X_C} = \frac{120 \text{ v}}{1.3 \times 10^3 \ \Omega} = 92 \text{ mA}$$

The rules for finding the total capacitance for capacitors in series and parallel are easy to remember: They are exactly the opposite of the rules for resistors in series and parallel.

$$C_{parallel} = C_1 + C_2 + C_3 + \ldots + C_n \qquad (3\text{-}4)$$

$$\frac{1}{C_{series}} = \frac{1}{C_1} + \frac{1}{C_2} + \frac{1}{C_3} + \ldots + \frac{1}{C_n} \qquad (3\text{-}5)$$

Figure 3-4 Circuit for use in Example 3-4.

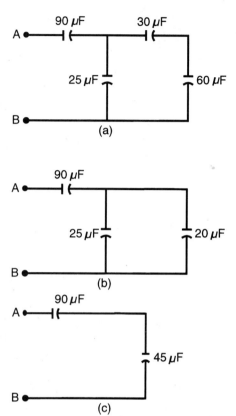

EXAMPLE 3-4

Find the capacitance between points A and B for the circuit shown in Figure 3-4a:
The first step would be to find the combined capacitance of the two capacitors in series, the 30 µF and the 60 µF capacitors,

$$\frac{1}{C_T} = \frac{1}{C_1} + \frac{1}{C_2}$$

or, better,

$$C_T = \frac{C_1 \times C_2}{C_1 + C_2}$$

$$= \frac{30 \ \mu F \times 60 \ \mu F}{30 \ \mu F + 60 \ \mu F}$$

$$= 20 \, \mu F$$

The circuit has now been reduced to that shown in **Figure 3-4b**. The next step would be to calculate the capacitance of the 25 µF and the 20 µF capacitors, which are in parallel.

$$C_T = C_1 + C_2$$

$$= 25 \, \mu F + 20 \, \mu F$$

$$= 45 \, \mu F$$

The circuit has now effectively been reduced to that shown in **Figure 3-4c**. The final step would be to calculate the total capacitance of the 90 µF capacitor in series with the 45 µF effective capacitance:

$$C_T = \frac{C_1 \times C_2}{C_1 + C_2}$$

$$= \frac{90 \, \mu F \times 45 \, \mu F}{90 \, \mu F + 45 \, \mu F}$$

$$= 30 \, \mu F$$

Time constant. Capacitors, combined with resistors, can be used to provide simple timing circuits, that is circuits that can regulate a dependent variable by providing a fixed or specified amount of time or time delay. This is possible because capacitors take a certain amount of time to charge and to discharge. This amount of time depends both upon the capacitance of the capacitor and the amount of resistance through which the current must flow to charge or discharge the capacitor. A simple example is shown in Figure 3-5.

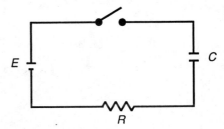

Figure 3-5 A capacitor-charging circuit.

When the switch is first closed, the capacitor begins to charge. The voltage across the capacitor is given by

$$E_c = E(1 - e^{-t/RC}) \tag{3-6}$$

where E_c is the voltage across the capacitor at any instant of time, E is the source voltage, and t is the amount of time (in seconds) since the switch was closed. This is shown in Figure 3-6.

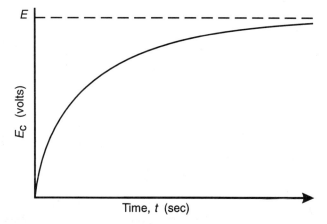

Figure 3-6 The transient voltage across a capacitor starting at the time the switch is closed.

Notice that at time $t = 0$, when the switch is first closed, Equation 3-6 gives $E_c = 0$, which makes sense because one would expect no voltage across the capacitor until charge has started to accumulate on the plates. At $t = \infty$, $E_c = E$, which is also logical because the capacitor continues to charge until it reaches the source voltage. Notice that in the steady-state condition (that is, after all effects of closing the switch have gone away), the voltage drop across the capacitor cancels the voltage provided by the source, thus satisfying Kirchhoff's Law.

The discharge of a capacitor as shown in Figure 3-7 when the switch is closed, is given by

$$E_c = E e^{-t/RC} \tag{3-7}$$

The quantity RC in Equation 3-6 and 3-7 is known as the *time constant* for each circuit, respectively, and has the units of seconds. The *RC time constant* is the amount of time that it will take the capacitor to charge to 63% of its final amount (or

Figure 3-7 The transient voltage across a discharging capacitor.

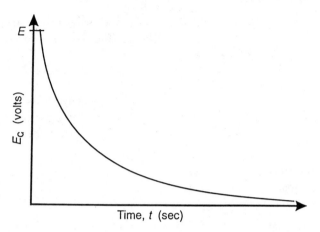

discharge 63% from its initial charge in the case of the second equation and circuit).

It is the time factor involved with charging and discharging a capacitor that makes it useful for inserting and controlling the variable *time* into a circuit such as a control circuit.

EXAMPLE 3-5

Consider the case where it is desired that a DC solenoid be activated one second after a switch is thrown (or, more realistically, after some other event has occurred). The solenoid is set up to obtain its driving voltage from across the capacitor shown in Figure 3-8. If a source voltage of 12 volts is used and the solenoid needs 10 volts to be activated, what value for R and C would be needed?

The procedure would be to solve Equation 3-6 for RC.

$$(1 - e^{-t/RC}) = \frac{E_C}{E}$$

Figure 3-8 A time-delayed solenoid.

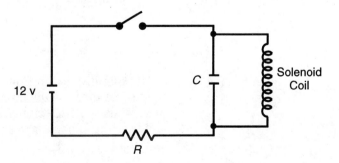

$$e^{-t/RC} = 1 - \frac{E_c}{E}$$

Taking the logarithm of both sides,

$$-t/RC = \ln(1 - \frac{E_c}{E})$$

$$RC = \frac{-t}{\ln[1 - (E_c \div E)]}$$

$$= \frac{-1}{\ln[1 - (10 \div 12)]}$$

$$= 0.558 \text{ sec}$$

Any combination of R and C which, when multiplied together, would give 0.558 would be a theoretical solution. (From a practical viewpoint, there would be other factors to consider, such as the fact that most capacitors are going to be on the order of µF or smaller. A one-farad capacitor could easily have the physical size of a five-gallon can.) For example, if a 1 µF capacitor were used, the resistor in the circuit would need to have a resistance of

$$R = \frac{0.558}{10^{-6}}$$

$$= 558{,}000 \text{ ohms}$$

$$= 558 \text{ kohms.}$$

Inductance

Another electrical characteristic that can be used for determining such variables as position is inductance. *Inductance* is a measure of the ease with which a voltage may be created in a device known as an *inductor*. Such voltages are created or *induced* by a magnetic field collapsing and rebuilding in the same region as the inductor. Magnetic fields have lines of force associated with them. As the magnetic field rebuilds and collapses, these lines cut across the electric wires of the inductor and induce voltages. The polarity of the voltages depends on the polarity of the field and whether it is building or collapsing. The electric wires in the inductor are usually arranged in the form of a coil so that the magnetic lines of force cut across the most lines practical for a given space. The size of the voltage created depends upon the rapidity with which the magnetic field lines change, the number of turns in the electric coil, and the strength and location of the magnetic field. The schematic symbol for an inductor is usually that of a coil, such as either side of the air core transformer shown in Figure 2-2.

The strength of the magnetic field in the area of the coil can be increased by placing an iron core in the center of the coil. Movement of this iron core into and out of the coil thus changes the effective inductance of the coil and—since inductance, like capacitance, can be measured by an electric circuit—there is once again the opportunity to measure position and related variables by measuring an electrical characteristic.

The effect of placing an iron core totally around a coil (as measured by a quantity known as *lines of force*) upon the strength of a magnetic field is shown in Figure 3-9.

It should be noted at this point that varying the placement of an iron core with respect to a magnetic field is more accurately known as varying the *reluctance* that the magnetic field sees. Magnetic field lines of force "travel" in a circuit. Opposition to their travel is referred to as *reluctance*. Reluctance is therefore related to magnetic field pathways as resistance is related to current pathways. When the iron core mentioned in the previous paragraph is moved, the reluctance is changed. Since the reluctance has an effect on the amount of magnetic field present, however, the net effect is that the inductance in the electrical circuit is changing. In sum, the more reluctance there is, the more "effective" inductance is seen by the inductance-measuring equipment. Although the reluctance is being varied by the iron core moving, the inductance is being directly measured.

The obstruction, X_L, of an inductor to AC current is dependent upon the inductance, L, of the coil or other inductor and the frequency, f, of the current, and is known as *inductive reactance*. It is also dependent upon whether resistors or capacitors are present in the circuit. Inductive reactance is measured in units of *henrys* (*H*) and expressed by Equation 3-8.

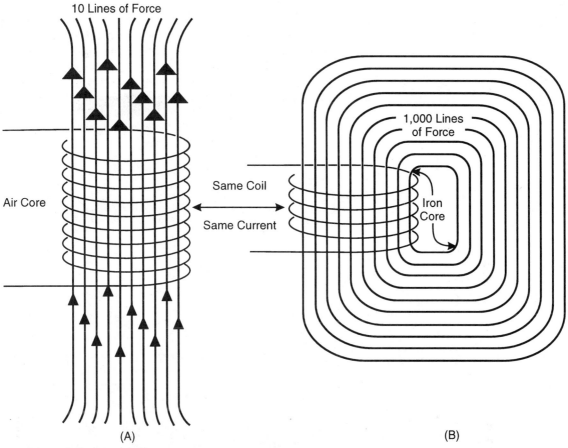

Figure 3-9 Effect of an iron core on the strength of a magnetic field.

$$X_L = 2\pi f L \qquad (3\text{-}8)$$

Inductive reactance has the same mathematical effect as resistance and capacitive reactance upon AC current (for a purely inductive circuit).

EXAMPLE 3-6

Compute the reactance of a 0.5 H inductor in series with a 2,000 cycles/sec AC voltage source.

$$X_L = 2\pi f L = 2\pi \cdot 60 \text{ cycles/sec} \cdot 0.5 \text{ H}$$

$$= 188 \; \Omega$$

Impedance

Inductance and capacitance have the unique relationship that they tend to cancel each other's impeding effect upon the flow of AC current. In fact, it has been found that the total *impedance*, Z, to AC current as a result of combinations of resistances, inductances, and capacitances is given

$$Z = \sqrt{R^2 + (X_L - X_C)^2} \qquad (3\text{-}9)$$

EXAMPLE 3-7

What is the amount of current flowing in the circuit shown in Figure 3-10?

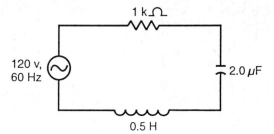

Figure 3-10 Circuit for use in Example 3-7.

From Example 3-5, it is known that $X_C = 1.3$ kΩ and from Example 3-6, $X_L = 188$ Ω, hence

$$Z = \sqrt{R^2 + (X_L - X_C)^2}$$
$$= \sqrt{1{,}000^2 + (188 - 1{,}300)^2}$$
$$= 1.5 \text{ k}\Omega$$

$$I = \frac{E}{Z} = \frac{120 \text{ v}}{1.5 \text{ k}\Omega}$$

$$= 0.079 \text{ A, or } 79 \text{ mA}$$

Phase Changes

When a capacitor initially begins to store electrical charge, there is no charge present on the plates, and current flows freely and easily onto the plates. As the amount of charge on the plates increases, the current must decrease because it is harder to place more and more electrons on the negative plate, which is already overly crowded with electrons. Likewise, it is harder to remove more electrons from the positive plate, which has already been severely depleted. In the first case, it can be said that when the voltage across the capacitor is zero, the current is at its highest. The second case is equivalent to saying that as the voltage across the capacitor increases, the current decreases. These cases are contained within the following formal rule: *The current leads the voltage in capacitive devices.* An alternative way of stating the same thing is to say the voltage lags the current in capacitive devices.

Because most alternating currents are sinusoidal in nature, the concept of *phase* is used to describe the extent to which the current leads the voltage in a capacitor. AC generators and other AC voltage sources typically produce an output electromotive force, or voltage, that can be defined by the sinusoidal Equation 3-10.

$$E(t) = E_m \sin(2\pi ft) \qquad (3\text{-}10)$$

In this equation, the E_m refers to the maximum amplitude of the voltage, the t to time, and the f to the frequency of the voltage source. When the generator is connected to a simple resistive circuit, as shown in Figure 3-1, the voltage across the resistor is shown in Figure 3-11, where it is plotted versus time. If the current through the resistor is calculated using Ohm's Law and then plotted versus time, it will appear as shown in Figure 3-12. In Figure 3-12, both the current and the voltage plots have been placed on the same graph. They both start at zero at the same time because when the voltage

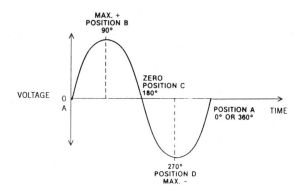

Figure 3-11 AC voltage across a resistor plotted versus time *(From Rex Miller and Fred W. Culpepper, Jr.,* Electricity and Electronics, *2nd ed., © 1991, Delmar Publishers, Inc., Albany, N.Y.)*

Figure 3-12 AC voltage and current across a resistor plotted versus time *(Adapted from Earl D. Gates,* Introduction to Electronics, *2nd ed., © 1991, Delmar Publishers, Inc., Albany, N.Y.)*

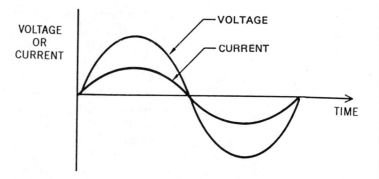

across the resistor is zero, the current through the resistor is zero. The voltage and current are said to be *in phase*.

If the same type of analysis is made for a capacitive circuit, the voltage and current plots will appear as in Figure 3-13. Notice from the plots that when the voltage is zero, the current is at its peak. When the voltage is at its peak, the current is zero. The voltage and current are said to be *out of phase*. They can also be said to have a *phase difference*. *Phase* is the fractional part of a cycle through which a periodic variable has advanced at any instant. From viewing the graphs, it can be seen that when the voltage is at its starting point, the current is already one-fourth of the way through its complete cycle. In other words, when the voltage phase is zero, the current phase is ¼. As was stated earlier, the current leads the voltage in a capacitive circuit.

Phase is usually measured in degrees or radians. A complete cycle is taken to represent 360° or 2π radians; hence zero phase is 0° (0 radians) and ¼ phase is 90° ($\pi/2$ radians). When the value of the phase is inserted into a mathematical equation, it is always in units of radians. The way phase is

Figure 3-13 AC voltage and current across a capacitor plotted versus time *(Adapted from Earl D. Gates,* Introduction to Electronics, *2nd ed., © 1991, Delmar Publishers, Inc., Albany, N.Y.)*

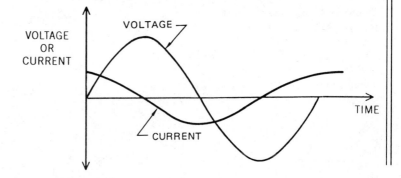

represented mathematically is shown by the general AC current equation, Equation 3-11.

$$I(t) = I_m \sin(2\pi ft - \phi) \tag{3-11}$$

I_m is the maximum peak of the current across the given device, f is the frequency of the AC source, and t is the time. The symbol ϕ, representing the phase difference between the current and the voltage, is measured with respect to the voltage. For the case of Figure 3-13, the voltage equation would be Equation 3-10 and the current equation would be Equation 3-11. For $t = 0$ and $\phi = 0$, the value of the voltage is zero, as it should be. For $t = 0$ and $\phi = -\pi/2$, the value of the current is at its maximum, as it should be. Both cases are now shown mathematically.

$$E(t) = E_m \sin(2\pi ft)$$

$$= E_m \sin(0)$$

$$= 0, \text{ for } t = 0.$$

$$I(t) = I_m \sin[2\pi ft - (-\pi/2)]$$

$$= I_m \sin(\pi/2)$$

$$= I_m, \text{ for } t = 0 \text{ and } \phi = -\pi/2$$

The best way to measure phase difference is from the maximum peak of the voltage to the nearest maximum peak of the current. Notice in Figure 3-14 that the phase difference is $-\pi/2$, even though the current is leading the voltage.

Figure 3-14 Example of how to measure phase difference.

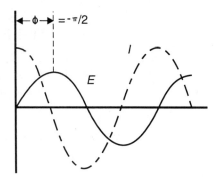

An inductor is just the opposite of a capacitor in regard to phase. *In an inductive circuit the current lags the voltage* by 90° or π/2. This is caused by the fact that an inductor tends to oppose any change of current passing through it. It opposes this change of current by creating a voltage across itself that opposes the change in the supply voltage. If the supply voltage is increasing, the inductor creates a voltage that opposes the supply voltage in an attempt to keep the current through it constant. For example, suppose the switch in Figure 3-15 is closed at the time that the AC source is at its negative peak. The inductor instantly imposes a voltage across itself that is exactly the opposite of the voltage supplied by the AC source at that same instant. Hence, there is an instantaneous voltage equal in value to the positive peak of the source across the inductor, and the current has not yet begun to build. This is probably the best scenario to keep in mind to remember that for an inductor the voltage leads the current.

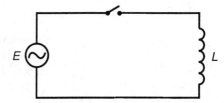

Figure 3-15 An AC inductive circuit.

Mathematically, the voltage and current across and through the inductor are expressed by Equations 3-12 and 3-13 when the AC source is a sinusoidal voltage.

$$E = E_m \sin(2\pi ft) \quad (3\text{-}12)$$

$$I = I_m \sin(2\pi ft - \pi/2) \quad (3\text{-}13)$$

EXAMPLE 3-8

Discuss the voltage/current phase relationship for the circuit shown in Figure 3-10:

The resistance in the circuit will do nothing to cause the voltage and current phases to noncoincide; however, the capacitance will cause the current to lead the voltage, and the inductance will cause the current to lag the voltage. Since the capacitive reactance is greater than the inductive reactance, the current will lead the voltage. (Exactly what portion of a cycle the current will lead the voltage can be calculated, but that subject is beyond the scope of this chapter.)

As mentioned previously, changes in capacitance and inductance are used to indirectly measure the change in physical variables. The phase-change phenomena just mentioned allows changes in capacitance and inductance to be measured. In addition, there are meters available with self-contained circuitry that will measure capacitance and inductance, just as there are meters available to measure voltage, current, and resistance.

Effective Current and Voltage

As mentioned earlier in this chapter, when the terms I and E have been used in the context of AC circuits, they have represented quantities known as *effective current* and *effective voltage*, respectively. When the instantaneous current is a sinusoidal function, as defined by Equation 3-11, the effective current is found to be 0.707 of I_m, the maximum peak of the current or, more explicitly,

$$I_{eff} = 0.707 I_m \qquad (3\text{-}14)$$

Likewise for the effective voltage:

$$E_{eff} = 0.707 E_m \qquad (3\text{-}15)$$

Effective current is the amount of AC current that effectively produces the same amount of heat or power (discussed in the next section) as would be produced by a numerically equal amount of DC current. The effective AC values are also the values that would be read by an AC ammeter or voltmeter. For example, when an AC voltmeter is connected to an electrical wall outlet, it will typically read a value on the order of 110 volts.

EXAMPLE 3-9

If an AC voltmeter is used to test a wall outlet in a lab and reads 110 volts, what is the peak-to-peak value of the AC voltage being tested?

From equation 3-15,

$$E_m = \frac{E_{eff}}{0.707} = \frac{110 \text{ v}}{0.707}$$

$$= 156 \text{ v}$$

That is, the AC line voltage is swinging from a high of 156 v to a low of −156 v in a sinusoidal waveform. If it is used to power some device such as an electrical motor, it will be able to supply the same amount of power as if it were a 110 v DC voltage source.

Hereafter, in AC situations, I and E shall be taken to mean I_{eff} and E_{eff}, respectively.

Electrical Power

Until now, the main electrical quantities that have been addressed are voltage and current. Another important quantity is power. For instrumentation purposes, *power* is the rate at which energy is consumed by a circuit. This energy may be consumed by heat being dissipated in a resistive device, by electromagnetic waves being generated by the circuit, or by other means. The unit of measurement for power is the *watt*. One watt is equal to one amp-volt.

The amount of power being supplied to any given device or circuit in a DC situation is found by multiplying the amount of current being supplied to that device by the amount of voltage drop across the device. In other words,

$$P = IE \qquad (3\text{-}16)$$

This equation also holds for an AC resistive circumstance where I and E are the effective values discussed in the preceding section. In addition, for any AC situation, the actual power being supplied to a device can be determined by

$$P = I_{eff} E_{eff} \cos\phi \qquad (3\text{-}17)$$

where ϕ is the phase difference between the current through and the voltage across the device and I_{eff} and E_{eff} are the maximum or peak values of the current and voltage. Notice that this latter equation is a very general equation that actually covers both AC and DC situations, since for DC the phase difference reverts to zero, and Equation 3-17 reverts to Equation 3-16. (For DC situations, the effective values of current and voltage would be the constant values.)

This formula holds true whether the current and voltage are DC or AC. The unit of power is *watts*. A watt is equal to a volt-amp.

EXAMPLE 3-10

How much power is dissipated as heat by the resistor in Figure 3-16?

Figure 3-16 Circuit for Example 3-10.

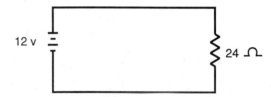

The current through the resistor would be given by Ohm's Law,

$$I = \frac{E}{R} = \frac{12 \text{ v}}{24 \text{ }\Omega} = 0.5 \text{ A}$$

Hence,

$$P = IE = (0.5 \text{ A})(12 \text{ v}) = 6 \text{ watts}$$

Use of Ohm's Law in Equation 3-16 gives another often-used power formula,

$$P = I^2 R \tag{3-18}$$

This equation is especially useful for showing why it is often advisable to transport electrical power using high voltage rather than high current. For a given resistance, for example the resistance in an electrical utility power line, the power expended as heat in the line is proportional to the square of the current.

EXAMPLE 3-11

It is required to transport 10,000 watts of electrical energy using a copper line with total resistance of 100 ohms. Would it be better to send the power to the other end of the line using high voltage or high current?

The first step would be to determine some trial values for the variables in question. Let high voltage be 10,000 volts. Then the current needed would be

$$I = \frac{P}{E} = \frac{10,000 \text{ w}}{10,000 \text{ v}}$$

$$= 1 \text{ A}$$

The power lost due to the resistance of the power line will be given by the following: (Note that the 10,000 watts is the power to be transported and should not be confused with the power that will be converted to heat energy in the line and lost. In fact, the amount of power to be sent needs to be equal to the amount to be transported to the other end plus the amount that will be lost in the line.)

$$P_{lost} = I^2R = (1 \text{ A})^2 \times 100 \text{ ohms}$$

$$= 100 \text{ watts}$$

This indicates that a total of 10,100 watts needs to be transmitted for 10,000 watts to be received for use at the far end. For the case of high current, assume a current of 10,000 amps is used. Then

$$P_{lost} = I^2R = (10,000 \text{ A})^2 \times 100 \text{ ohms}$$

$$= 10^{10} \text{ watts,}$$

which is a considerable difference. It is obvious that electrical power is best transported in high voltage form.

Transformers

One advantage of AC electricity is that it can be used to transfer power (or voltage or current) from one circuit to another without any direct connection between the circuits. A device

Figure 3-17 Schematic symbol for transformer *(From Rex Miller and Fred W. Culpepper, Jr., Electricity and Electronics, 2nd ed., © 1991, Delmar Publishers, Inc., Albany, N.Y.)*

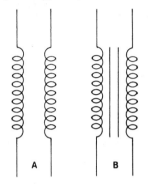

Figure 3-18 Magnetic field of a transformer *(From Rex Miller and Fred W. Culpepper, Jr., Electricity and Electronics, 2nd ed., © 1991, Delmar Publishers, Inc., Albany, N.Y.)*

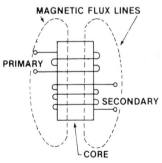

that can transfer electrical power from one coil to another coil is known as a *transformer*. The schematic symbol for this device is shown in Figure 3-17 (with an air core on the left and an iron core on the right), and also in Figure 2-2. Transformers typically consist of two separate coils of wire wound on the same core.

Transformers use the phenomenon of induction to perform their tasks. As an increasing voltage in a primary circuit causes current to enter a transformer's primary coil, a magnetic field is built about the coil, as shown in Figure 3-18. The increasing or building magnetic field lines from the primary coil cut across the individual turns in the secondary coil, inducing a voltage across the secondary coil. The relationship between the voltage across the secondary coil and the voltage across the primary coil depends on the relative number of turns in the coils, as expressed by Equation 3-19:

$$\frac{V_s}{V_p} = \frac{N_s}{N_p} \qquad (3\text{-}19)$$

In this equation, N_s stands for the number of turns of wire in the secondary coil and N_p is the number of turns in the primary coil.

Transformers are very commonly used to step voltages down. For example, an incoming line voltage of 110 volts AC might be stepped down to 12 volts AC by a transformer in an instrument so that the lower voltage would be available for use within the device. A rectifier would probably be available to convert the low AC voltage to an even lower DC voltage. Low voltages are preferred within instruments both to prevent electrical arcing between parts and to ensure safety.

EXAMPLE 3-12

If the primary had 1,200 turns, for the case just discussed, how many turns would the secondary need to step the voltage down to 12 volt AC?

$$\frac{V_s}{V_p} = \frac{N_s}{N_p}$$

$$N_s = \frac{V_s \cdot N_p}{V_p} = \frac{12 \text{ v} \cdot 1{,}200 \text{ turns}}{110 \text{ v}}$$

$$= 131 \text{ turns}$$

Iron cores are often used within transformers to make the transfer of power more efficient. Ferromagnetic materials such as iron tend to attract magnetic field lines, thus concentrating them in certain areas. An iron core within a transformer tends to increase the number of internal field lines within the two coils and thus encourages the maximum amount of cutting of wire turns with field lines.

Electromagnetism

Magnetic fields have been mentioned several times in earlier sections of this chapter; hence, a slightly more detailed discussion of magnetism and electromagnetism is in order. The two subjects are related in the sense that *electromagnetism* is magnetism caused by an electric current. The lines of force identified with a magnetic field are more formally known as *magnetic flux*.

Electricity and magnetism are intimately related because each can cause the other. Current flowing through a wire creates a magnetic field about the wire, as shown in Figure 3-19.

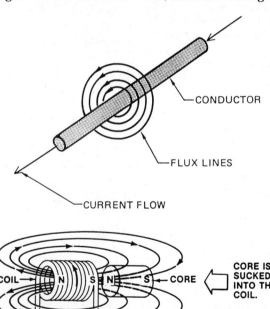

Figure 3-19 Magnetic field created by a current flowing in a conductor *(Adapted from Earl D. Gates,* Introduction to Electronics, *2nd ed., © 1991, Delmar Publishers, Inc., Albany, N.Y.)*

Figure 3-20 Electromagnetic action of a solenoid *(Adapted from Rex Miller and Fred W. Culpepper, Jr.,* Electricity and Electronics, *2nd ed., © 1991, Delmar Publishers, Inc., Albany, N.Y.)*

Alternatively, a changing magnetic field in the vicinity of an electrical conductor causes a current in the conductor. Both of these phenomena are extremely useful for implementing electromagnetic devices such as solenoids, motors, relays, and others. The use of electromagnetism to create a solenoid is shown in Figure 3-20.

REVIEW MATERIALS

Important Terms

hertz
reactance
inductance
phase
phase difference
transformer
frequency
effective current
time constant

capacitor
impedance
reluctance
AC voltage
electrical power
solenoid
period
effective voltage
electromagnetism

Questions

1. Why would one not see very many capacitors in DC circuits but see numerous capacitors in AC circuits?
2. Why does the addition rule for capacitors in parallel make sense in a physical/electrical way?
3. In what way are a capacitor and an inductor related to a resistor?
4. If the frequency of an AC source is increased, does a capacitor become more or less of an impedance?
5. Answer Questions 2–4 for the case of an inductor.
6. Suppose one doubles the plate size of a capacitor. Does it now present more impedance or less impedance to AC current?
7. List some materials that could be used as capacitor dielectrics.
8. Why would some dielectrics function better than others?
9. How is variable capacitance related to instrumentation?
10. What are three easy ways to vary capacitance?
11. How does reluctance differ from resistance?
12. If a coil had its number of turns decreased by ½, would its inductance increase or decrease?
13. Why could a transformer be viewed as an "isolation" device?

14. *Phase* represents the present portion of a cycle of an AC voltage or current. True or false? Discuss.
15. Why is the effective AC voltage used in place of the peak AC voltage?
16. What are various ways in which power can be used or dissipated by electrical devices?

Problems

1. What capacitive reactance would a 1.5 picofarad capacitor present to a 60 Hz source? To a 60,000 Hz source? (Pico- means 10^{-12}.)
2. For a certain instrumentation setup it is desired to have a capacitance available of at least 1.0 µF at all times. Occasionally the plates may be as far apart as 0.5 meter. The plates will be in vacuum (see Example 3-2). What is the minimum acceptable area of the plates?
3. Calculate the amount of current flowing in the circuit shown in Figure 3-21.

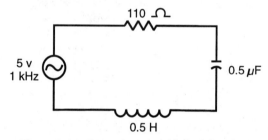

Figure 3-21 Figure for use with Problem 3.

4. Is the total reactance in Problem 3 more capacitive or more inductive? Sketch a rough graph of the source voltage and the current for the circuit in Figure 3-21, showing only a relative phase difference between the current and voltage. (In other words, show the current either leading or lagging the voltage.)
5. The current lags the voltage by $\pi/4$ in a television set. The set consumes 250 watts of power. Assuming the effective line voltage is 110 v, calculate the current flowing in the power cord to the set.
6. Suppose one wishes to reduce the AC voltage being supplied to an instrument from 24 v to 6 v. What would the ratio of the number of turns in the secondary coil to the number of turns in the primary coil need to be?
7. In Figure 3-22 there is a cross-sectional view of a conductor facing toward (or away from) you. The current in the conductor is flowing away from you (into the paper, so to speak). Draw the magnetic lines of flux associated with this current flow, using arrows to indicate the direction of the lines of flux.
8. Assume that a given transformer is 100% efficient; that is, the power being supplied by the primary side is transported without loss to the secondary side. (Transformers are *not* 100% efficient. One can detect this by the heat that they radiate.) The primary side has 100 turns. If the current and voltage on the primary side are 1 amp and 10 volts, respectively, how much current is flowing on the secondary side?

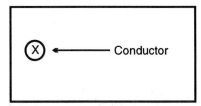

Figure 3-22 Figure for use with Problem 7.

8. Assume that a given transformer is 100% efficient; that is, the power being supplied by the primary side is transported without loss to the secondary side. (Transformers are *not* 100% efficient. One can detect this by the heat that they radiate.) The primary side has 100 turns. If the current and voltage on the primary side are 1 amp and 10 volts, respectively, how much current is flowing on the secondary side?

Electronics

CHAPTER GOALS

After study of this chapter, you should be able to do the following:

- Define the term *electronics*.
- Identify electrical components as either active or passive.
- State what analog electronics is and how it differs from digital electronics.
- Know the characteristics and uses of an operational amplifier.
- Understand the concepts of feedback and signal inverting, especially as applied to operational amplifiers.
- State the basic working principles and adjustment techniques of a regular amplifier.
- Define the concept of gain.
- Explain how a DC amplifier differs from what is normally known as simply an amplifier.
- Discuss the use of the following special-purpose amplifiers: differential amplifiers, buffer amplifiers, current amplifiers.
- Know the potential effect of a low-input impedance of a circuit on the output of the preceding circuit.

BASIC CONSIDERATIONS

Many electrical devices are made of semiconductor materials. The study of these devices and the circuits that incorporate them is known as *electronics*. Electronics also includes the study of electrical circuits involving vacuum tubes, discharge tubes, and other similar devices; but semiconductor devices are by far the most common.

Most of these electronics devices are known as *active* devices because they amplify or control. Their opposites are the *passive* components mentioned earlier in Chapters 2 and 3—the capacitor, inductor, resistor, and any other component that does not amplify or control.

ANALOG ELECTRONICS

Electronics can be divided many different ways but one distinct division is analog electronics and digital electronics. *Analog electronics* is the study of electronic circuits where the output is a continuously varying function of the input. The most common example is the simple electronic amplifier: When the input doubles, the output doubles.

Operational Amplifiers

The best known of the semiconductor devices are the transistor and the integrated circuit. An *integrated circuit* is a circuit composed of a multitude of active semiconductor devices such as transistors and passive semiconductor devices such as resistors and capacitors that is carved from a single microscopic piece of semiconductor material commonly known as a *chip*. A special integrated circuit, known as the operational amplifier, is discussed in this section of the textbook because of its wide use in instrumentation and because it provides a sensible introduction to the use of electronics in instrumentation.

The *operational amplifier*, more commonly known as the *op-amp*, is a relatively small integrated circuit used to amplify electrical signals. Most integrated circuits are built in dual inline pin package (dip) form, as shown in Figure 4-1, and most have eight pins on each side. Some, such as a microprocessor, have many more than this, however, while some have only four pins on each side. The operational amplifier is usually of this latter form. Figure 4-2 shows the internal details of the most common op-amp, the 741.

Figure 4-1 Two common IC packages *(From Rex Miller and Fred W. Culpepper, Jr., Electricity and Electronics, 2nd ed., © 1991, Delmar Publishers, Inc., Albany, N.Y.)*

National Semiconductor

Operational Amplifiers/Buffers

LM741/LM741A/LM741C/LM741E Operational Amplifier

General Description

The LM741 series are general purpose operational amplifiers which feature improved performance over industry standards like the LM709. They are direct, plug-in replacements for the 709C, LM201, MC1439 and 748 in most applications.

The amplifiers offer many features which make their application nearly foolproof: overload protection on the input and output, no latch-up when the common mode range is exceeded, as well as freedom from oscillations.

The LM741C/LM741E are identical to the LM741/LM741A except that the LM741C/LM741E have their performance guaranteed over a 0°C to +70°C temperature range, instead of −55°C to +125°C.

Schematic and Connection Diagrams (Top Views)

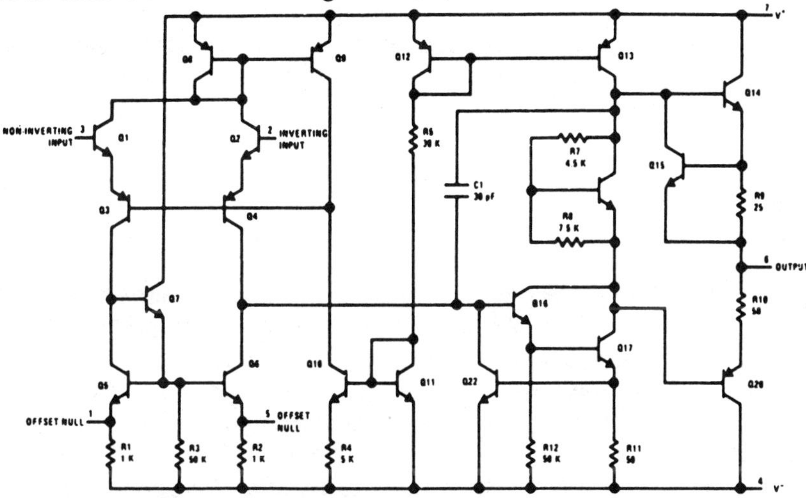

Metal Can Package

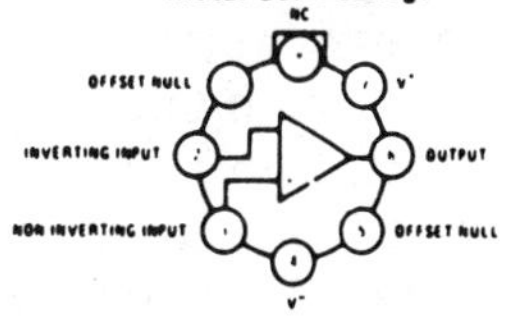

Order Number LM741H, LM741AH, LM741CH or LM741EH
See NS Package H08C

Dual-In-Line Package

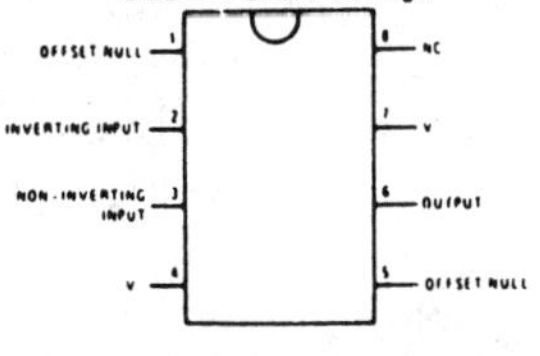

Order Number LM741CN or LM741EN
See NS Package N08B
Order Number LM741CJ
See NS Package J08A

Dual-In-Line Package

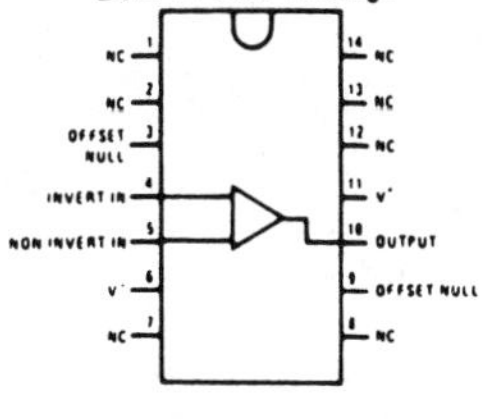

Order Number LM741CN-14
See NS Package N14A
Order Number LM741J-14, LM741AJ-14
or LM741CJ-14
See NS Package J14A

Figure 4-2 Detailed description of the 741, a very common op-amp *(From Earl D. Gates, Introduction to Electronics, 2nd ed., © 1991, Delmar Publishers, Inc., Albany, N.Y.)*

The schematic symbol for an operational amplifier is shown in Figure 4-3. This indicates that only three of its eight pins are usually relevant to circuit operation. The other pins are either dummy pins or are used for functions that are not germane to understanding the design of a functional circuit. For example, power to the operational amplifier is always supplied through two pins, and this never varies. This is not information that needs to be shown on the schematic. (A unique feature of operational amplifiers is that they always require a ±-type power supply, typically on the order of ±15 volts.) The resistors and other passive devices surrounding the op-amp are definitely items of interest, however, because the way they are interconnected with the op-amp determines the function the op-amp is to play.

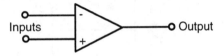

Figure 4-3 Schematic symbol of an op-amp.

Figure 4-4 shows an op-amp being used as an amplifier. The resistor connecting the output back to the input provides *feedback*. A part of the output signal is fed back to the input where it cancels part of the input signal because of the opposite polarities of the two signals. The resistor is present in the feedback line to prevent all the output signal from being fed back to the input. Since the input signal is less than the output signal, a condition true for nearly all amplifiers by definition, only a very small part of the output signal needs to be sent back to the input. Op-amps without feedback have very large amplification factors—on the order of 10^5. Because of this huge amplification factor, they go into runaway oscillation if there is no feedback. With feedback, as the output signal tries to get larger and larger, the feedback signal gets larger and larger, canceling even more of the input signal and shutting down the runaway oscillation. Op-amps cannot operate in any circuit without some type of feedback to keep them under control. By choosing the size of the feedback resistor, one can choose the amplification factor of the circuit.

When the input to the operational amplifier is fed to the negative (–) terminal of the op-amp, as shown in Figure 4-4, the output will be inverted. A signal is said to be *inverted* when the polarity of the signal is reversed, that is, where all negative parts of the signal are changed to positive and vice

Figure 4-4 Op-amp amplifier circuit with inverted output *(From Earl D. Gates, Introduction to Electronics, 2nd ed., © 1991, Delmar Publishers, Inc., Albany, N.Y.)*

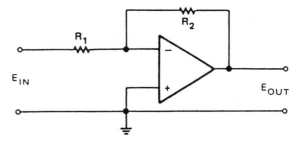

versa. If the signal is fed to the positive (+) terminal of the op-amp, the output is not inverted. This is shown in Figure 4-5.

Operational amplifiers are to some extent the all-purpose workhorses of the electronics family of integrated circuits. They can be used to build simple amplifier circuits, but they can also be used for many other purposes: specialized amplifiers such as differential amplifiers and buffer amplifiers, filters, adders, subtracters, differentiators, integrators, and so on. Some of these uses will be looked at now and some later in the textbook.

Figure 4-5 Op-amp amplifier circuit with noninverted output *(From Earl D. Gates, Introduction to Electronics, 2nd ed., © 1991, Delmar Publishers, Inc., Albany, N.Y.)*

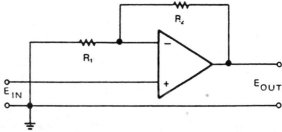

Regular Amplifiers

Regular amplifiers perform no unusual function: They simply amplify an incoming voltage or signal and provide it as an output voltage or signal. They are called *amplifiers*, with no descriptive adjective in front of the word. The term *regular amplifier* was used here merely to distinguish the classification from the more special-purpose amplifier classifications to follow.

Amplifiers vary in the number of controls available to manipulate their performance. Some have no controls (such as adjusting knobs or switches) available. They have fixed amplification such as "×100". A ×100 amplifier produces an output voltage that is 100 times greater than the input voltage (provided that this output does not exceed the supply voltage used to power the amplifier; in general, an amplifier cannot provide an output greater than the internal power voltage that runs it).

Amplifiers exist mainly to provide gain. *Gain* is what heretofore in this chapter has been called the amplification factor, or simply amplification. It is most commonly the output voltage divided by the input voltage, as given by the following equation:

$$\text{gain} = \frac{\text{output voltage}}{\text{input voltage}} \qquad (4\text{-}1)$$

Gain need not involve only voltage. There are other types of amplifiers for explicitly amplifying current or power. The more generalized equation for gain, then, is given by

$$\text{gain} = \frac{\text{output}}{\text{input}} \qquad (4\text{-}2)$$

One should note that in the use of Equation 4-2 the output and input are required to have the same units. If they do not have the same units, then one does not refer to the gain of the circuit but instead to its transfer function or transfer ratio. This is discussed at greater length both in the appropriate sections of this chapter and in Chapter 6.

Some amplifiers have adjustable gain controls. One common configuration is to have a coarse gain control and a fine gain control. Such an example is shown in Figure 4-6. Here, there is a switch to change the coarse gain from ×1 to ×10 to ×100. Within each of these ranges, the gain may be fine-tuned with the fine gain control, which allows the operator to continuously further adjust the coarse gain just selected. For example, if the operator selects ×10 as the coarse gain setting, the fine gain control can be used to change the overall amplification factor to anywhere between ×10 and ×100. The total amplification is approximately the product of the two gain settings.

EXAMPLE 4-1

Suppose the two gain settings of the amplifier shown in Figure 4-6 are as shown. What is the approximate amplification set-

Figure 4-6 A typical adjustable amplifier.

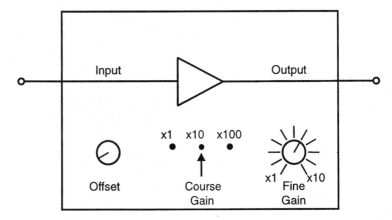

ting of the amplifier? (It would be necessary to determine the actual amplification by measuring the output voltage and dividing it by the input voltage.)

With the coarse gain setting at x10, the amplification will be between x10 and x100. The fine gain control is approximately at 6.5; hence, the total gain is 10 x 6.5 or 65. Note that this is between 10 and 100 as predicted.

Many amplifiers have what is known as an *offset* control. This adjustment is to assure that the amplifier produces zero output when it has zero input. Stray voltages can make their way into amplifiers and produce output when none is expected or give erroneous output when an output *is* expected. For example, most op-amps use power-supply voltages of ±15 volts. If one of the supplies is off just 1 volt (say, the negative supply is giving −14 v instead of −15 v), then the amplifier will not necessarily provide 0 v output for 0 v input. To compensate for this, the offset control is used. If the amplifier has only one input, this input is grounded and the offset control adjusted until there is zero output. If the amplifier has two inputs (as in the case of the differential amplifier discussed later), then the two inputs are shorted by connecting them with a wire and the offset control is used to adjust the output to zero volts.

DC Amplifiers

Most amplifier circuits have a capacitor in the input lead, although this is usually not shown in the simple schematic diagrams in this book. The purpose of the capacitor is to block any superfluous DC signal voltage that might be present while allowing the AC signal voltage to pass through to the amplifier. This is a common set-up because it is AC signals that are usually amplified, not DC signals. More to the point, most signals are AC, not DC. There are cases where signals may be DC, however; so when an amplifier is built without a capacitor in the input (and, likewise, without a capacitor in the output), it is denoted as a *DC amplifier*. AC amplifiers are most commonly called simply *amplifiers*. Since capacitors are present in most amplifier inputs and outputs, they are usually not shown on a brief diagram because it is assumed that they are there.

Differential Amplifiers

A *differential amplifier* is basically a *subtracter*. It is an amplifier with two inputs. The output is a function of the voltage difference between the two inputs. Typically, differential amplifiers have a gain of 1; that is, the output simply equals the difference between the two inputs, with no amplification of this difference. They are also available with amplification, such as fixed amplification of 100 or adjustable amplification. If the amplifier is not supplied with adjustable gain or offset control, this can easily be supplied by feeding the output of the differential amplifier into the input of an ordinary adjustable amplifier.

EXAMPLE 4-2

Suppose one has an amplifier set-up as shown in Figure 4-7, where a differential amplifier serves as the input and feeds its output into the input of an adjustable amplifier. What steps would one take to ensure that a final output of 12 v is obtained when the two inputs are 150 mv and 90 mv, respectively?

The first step would be to set the gain as low as possible and then to work with the offset. The coarse gain switch would be set to ×1. The fine gain setting would be set counterclockwise as far as possible (so that it is basically in its ×1 position). Next, the input of the differential amplifier would be shorted by connecting a wire lead between the two inputs. Finally, the offset control would be adjusted until the second amplifier shows zero volts output. The entire amplifier setup now has its output zeroed for ×1 gain and zero input. This procedure should now be

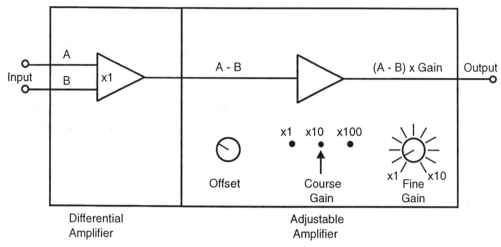

Figure 4-7 An adjustable differential amplifier set-up.

repeated with the gain set to x10 (coarse control only, fine control should remain at x1) and then with the gain at x100.

The next step would be to set the gain controls to provide the gain sought. In this case, this would be given by the following:

$$\text{gain} = \frac{\text{output voltage}}{\text{input voltage}}$$

$$= \frac{12 \text{ v}}{0.15 \text{ v} - 0.12 \text{ v}}$$

$$= 400$$

Thus, the coarse gain control would be set to x100 and the fine gain control to 4. If the two inputs were known to be 150 mv and 120 mv, then the fine gain control could be used to "fine tune" the output to 12 v if it was not quite 12 v. Some other known inputs could also be used to adjust the gain until it was as close to 400 as possible. These inputs would need to be fed into the differential amplifier, however, since it is the entire amplifier set-up that is to be adjusted.

The opposite device from differential amplifiers are called *adders*. These devices add the two inputs instead of taking the difference, and output the resulting sum.

Current Amplifiers

The term *current amplifier* has two different meanings in the world of industrial instrumentation. The traditional electronics definition of this term is that it is a device that amplifies current; that is, it takes an input current, amplifies it, and outputs this amplified current. In the area of industrial instrumentation and control, however, it is often another name given to an amplifier circuit that is a special case of a class of devices known as current-to-voltage converters. This class of devices is discussed in general in Chapter 6, Signal Transmission. A current amplifier by this latter definition has an input of current and an output of voltage. The ratio between the output and the input is known as the transfer ratio or transfer function. (It cannot correctly be called gain, since the output and input have different units.) This is symbolized by Equation 4-3:

$$\text{transfer ratio} = \frac{\text{output voltage}}{\text{input current}} \tag{4-3}$$

EXAMPLE 4-3

For the example shown in Figure 4-8, what is the transfer ratio?

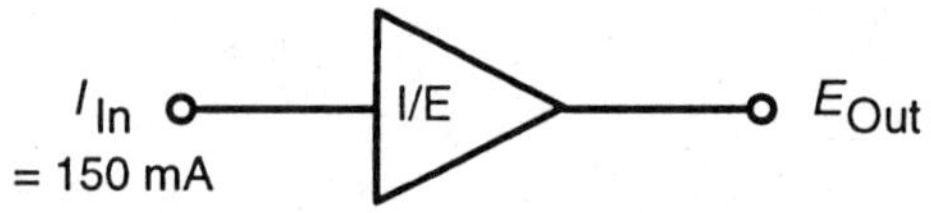

Figure 4-8 Schematic for use with Example 4-3.

$I_{In} = 150$ mA, I/E, E_{Out}

The input current is 0.15 A and the output voltage is 1 v. The transfer ratio is then

$$\text{transfer ratio} = \frac{\text{output voltage}}{\text{input current}}$$

$$= \frac{1 \text{ v}}{0.15 \text{ A}}$$

$$= 6.67 \text{ v/A}$$

Buffer Amplifiers

Buffer amplifiers are a type of impedance-matching devices. Every circuit has an input impedance and an output impedance. Impedance is the AC equivalent to DC resistance. A DC current sees impedance to its flow only in the form of resistance. AC current, on the other hand, sees impedance to its flow not only due to the inherent resistance of the conductors, but also due to capacitors and inductors. Power is best transferred from one circuit to another if the input impedance of the second circuit is about equal to, or "matched by," the output impedance of the first circuit.

It is also desirable—all other factors being equal—for the input impedance of a circuit to be high, or at least of medium impedance, in a relative sense. If the input impedance of a circuit is low, it can "load" the circuit immediately in front, causing the first circuit to output results other than those expected.

EXAMPLE 4-4

Consider the 10,000 ohm potentiometer in Figure 4-9a that has 10 volts across it. This potentiometer is being used to supply a variable voltage to a second circuit immediately following it. The second circuit is an amplifier with an input impedance of 1,000 ohms. Not considering the purpose of the overall set-up, is this a workable arrangement?

Assume the potentiometer is at its halfway position. The expected output from the pot is 5 v. However, the actual circuit seen by the current is shown in Figure 4-9b. Notice that the output voltage is much less than expected, an actual 1.4 volts versus an expected 5 volts. The formal answer to the question posed, then, is "No." The set-up is not acceptable because the first circuit had an output of 5 volts, which was to be amplified by the second circuit. Instead of working with the output of the first circuit, however, the second circuit actually modified the operation of the first circuit.

The answer to the problem posed by Example 4-4 is the use of a buffer amplifier between the two circuits. A buffer amplifier has a high-input impedance, often on the order of 100 kohms. This input impedance is large enough to have an unnoticeable effect on the preceding circuit. They are also designed to have a

Figure 4-9 Schematic for use with Example 4-4.

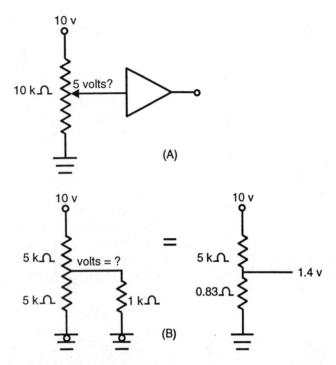

low-output impedance (say, on the order of 1 kohm) so as to be compatible with almost any circuit that is to immediately follow. Typically, they have a gain of 1. In other words, they are not used as amplifiers. They simply serve to allow a high-impedance output circuit to be matched to a low-impedance input circuit. The buffer amplifier fits between the two circuits and does nothing but keep the second circuit from affecting the first.

EXAMPLE 4-5

What would be the actual "loading" effect of the typical buffer amplifier with an input impedance of 100 kohms on the output of the potentiometer mentioned in Example 4-4?

The actual circuit seen by the output of the potentiometer is shown in Figure 4-10.

DIGITAL ELECTRONICS

Digital electronics is the study of electronic circuits where the input values are discrete, and these discrete inputs produce discrete outputs. Most digital electronic circuits limit

Figure 4-10 Schematic for use with Example 4-5.

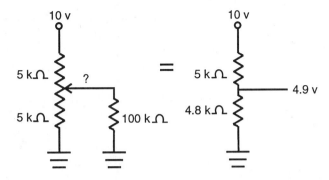

these discrete inputs and outputs to two small ranges of values. The main thing to recognize is that analog refers to continuous values and digital refers to discrete values.

Digital techniques are used in electronics and, in particular, in computers for the reason of accuracy. Digital voltage signals are less easily compromised or degraded than are analog signals.

As mentioned previously, all digital voltage signals are pulses. Basically, these pulses are supposed to have only two voltage values, typically called *high* and *low*, *up* and *down*, or *on* and *off*. To allow for degradation of the pulse or signal, however, the pulses are allowed to take on values within a specified range and still be considered as high or low. One common set of nominal values is +5 v for the high signal and 0 v for the low signal. Any pulse within the range +2.4 to +5.0 v might perhaps be taken to be a high signal, with 0.0 to +0.5 v being the range for a low signal. Any pulses outside these ranges are taken to be erroneous and are ignored.

These pulses are used to transmit information from one part of the circuit to another, from one circuit to another, or from one piece of equipment to another. To send the smallest amount of information (e.g., just the value "7") takes at least eight pulses, usually transmitted simultaneously through parallel pathways. These pulses, which together represent information, are individually referred to as *bits*, a shortened name for binary digit. The word *binary* is used because there are only two values the pulse or bit can assume, which is associated in turn with the fact that digital electronics is operated using the base-2 number system.

Digital electronics works by the brute force of sheer numbers. It takes many more electrical pulses to convey information than analog electronics, but there is very little delay because all electrical pulses travel very fast (theoretically, at the speed of light) and they are usually sent simultaneously through parallel channels. To send the number "7" from one location to another can easily take 32 pulses or bits in a digital system where it would take only one 7 v signal or pulse in an analog system. In a digital system, however, the receiving circuit knows to expect only incoming signals that are within the small acceptable ranges for high and low pulses; and if anything else is received, an error is known to have occurred. The system then asks for retransmission of the information, or it displays an error message. In an analog system, the receiving circuit accepts whatever it receives. If the 7 v signal were degraded to, say, 6.9 v due to resistance in the circuit wires, the receiving circuit would not know about the error. In sum, digital systems use only two values for all pulses for purposes of accuracy.

Most, if not all, digital electronics in current times is semiconductor in nature. The various semiconductor devices include gates (AND, OR, NAND, NOR, etc.), inverters, memory chips, microprocessors, analog-to-digital converters (A/D or ADC), and digital-to-analog converters (D/A or DAC). The only digital devices that will be discussed to any length in this book are the ADC and the DAC, and this discussion is postponed until Chapter 6.

REVIEW MATERIALS

Important Terms

electronics
active components
analog
integrated circuit
op-amp
inverter
gain
DC amplifier
adder
current amplifier
input impedance
buffer amplifier
high/low
binary

operational amplifier
passive components
digital
chip
feedback
amplifier
offset
differential amplifier
subtracter
transfer ratio
output impedance
loading
bit
pulse

Questions

1. List as many active electrical devices and as many passive electrical devices as you can think of.
2. If the input signal is connected to the noninverting (+) input of an op-amp, is feedback still needed?
3. Assume an op-amp is an 8-pin IC. What would be the uses of the 8 pins?
4. What would be the highest voltage output for the typical op-amp? Does it depend on the op-amp's gain?
5. How could one modify the op-amp circuit shown in Figure 4-4 to make it an AC only amplifier?
6. Suppose the feedback resistor in Figure 4-4 was 2,000 ohms. If it were replaced by a 1 kohm resistor, would the amplification be increased or decreased? Why?
7. If a differential amplifier has no offset control, can one arrange a set-up so that one can be sure the differential amplifier has zero output when the two inputs are short-circuited?
8. If an amplifier is found that always produces 50 times as much output current as input current, is it a current amplifier?
9. If the voltage level standard discussed in this chapter for high bits and low bits were being used, would a 1.0 v voltage pulse be acceptable? A −1.0 v voltage pulse?
10. If you had to use only one word to describe why digital is oftentimes preferred to analog in the electronics world, what would it be?
11. If one had an op-amp creating signals that were being sent to a computer, which would you use—an ADC or a DAC?
12. Could one also adjust the offset control of a dual differential-amplifier/adjustable-amplifier set-up as shown in Figure 4-7 by connecting each of the two inputs to ground instead of short-circuiting them with one another?
13. Why would one adjust the offset of an amplifier with the gain set to the lowest value first, then readjust as the gain is increased? See Example 4-2.

Problems

1. Draw a schematic of an op-amp being used as an noninverting amplifier *without feedback*. Will a set-up such as this actually function in a useful manner?
2. Draw a schematic showing an op-amp circuit being used as an amplifier with adjustable gain (both fine and coarse). Hint: Use fixed resistors for the "x1, x10, x100" switch and a pot for the fine gain control.
3. How might one add the offset control to the circuit of Problem 2?
4. Design an op-amp circuit so that it acts as an amplifier for the DC portion of an input but not for the AC portion of an input. Note: This problem is the exact opposite case of Question 5 but requires more thought.

5. Assume an op-amp circuit is set up as an inverting amplifier with a gain of 100 and that the input signal is a 60 Hz, 5 v peak-to-peak sinusoidal voltage. Sketch a graph of one cycle of both the input signal and the output signal.
6. Draw the schematic for two op-amps being used to make an adder.
7. If a differential amplifier has inputs of 0.15 mv and 0.09 mv and an output of 4.5 v, what is its gain? What is its transfer ratio?
8. What would be the gain control settings for an amplifier with coarse and fine gain controls (as discussed in this chapter) to obtain a total gain of 377? A gain of 7.5?
9. What is the gain of a current-to-voltage converter that produces a 10 v output with a 0.220 mA input? The transfer ratio?
10. If a differential amplifier had an output impedance of 60,000 ohms and was to be used in a dual amplifier set-up with the differential amplifier as the first amplifier and an adjustable amplifier (input impedance of 100 kohms) as the second amplifier, would a buffer amplifier need to be used between them? If so, what impedance characteristics would it need to have?

Pressure

CHAPTER GOALS

After completing study of this chapter, you should be able to do the following:

Understand the meaning of the term *pressure*.

Calculate various pressures depending on the information given.

Understand the difference between atmospheric, absolute, and gage pressures.

Use the concept of "differential pressure."

Use the various units with which pressure is given (psi, Pa, etc.).

Apply the concepts of hydrostatic pressure and specific weight.

Understand the meanings of static, dynamic, and total or impact pressure.

Understand the hydrostatic paradox.

Work with the concept of vacuum or negative pressures.

Understand where a body's buoyancy comes from and how to determine it.

Understand and apply Pascal's Law.

Describe how various manometers, gages, and transducers measure pressure and the situations where each can be used.

Understand how a barometer works.

Apply what you learn to selecting, installing, calibrating, and protecting pressure-measuring devices.

PRESSURE

Pressure cannot be "seen," but its effects are part of our daily activities. We live at the bottom of an atmosphere of air. Water is one of the keys to life on earth. The pressures involved with air, water, and other fluids play an important part in industry through their effect on boiling and condensing points, hydraulic and pneumatic equipment, process efficiency, cost, and other items. Pressure can also be used in the measurement of density, flow, liquid level, and temperature.

Pressure measurements are usually given as a force acting over an area or in terms of a column of liquid. Devices for sensing pressure include manometers, gages, and transducers. When setting up a pressure measuring system, consideration should be given to device selection, installation, calibration, and protection.

BASIC CONSIDERATIONS

Pressure is the measure of a force acting over an area. It is calculated by dividing the magnitude (size) of the force by the area over which the force acts:

$$\text{pressure} = \frac{\text{force}}{\text{area}}$$

Pressure is commonly expressed in pounds per square inch or psi.

EXAMPLE 5-1

A force of 300 lbs acts over an area of one square foot. What is the equivalent pressure?

$$\text{pressure} = \frac{300 \text{ lbs}}{144 \text{ in}^2} = 2.08 \text{ psi}$$

Basic Terms

Atmospheric and Absolute Pressure. The weight of the air in the earth's atmosphere causes a pressure on objects within the atmosphere. This pressure is called *atmospheric* and varies with distance above sea level and with weather conditions. Sea level atmospheric pressure is normally given as 14.7 psi; at an elevation of 5,000 feet, it drops off to 12.2 psi. Because any pressure measured from a reference or datum of zero pressure is called an *absolute pressure*, the atmospheric pressures as reported in newspapers or weather reports are absolute values.

EXAMPLE 5-2

A pressure gage reads 10 psi. What is the absolute pressure if the atmospheric pressure is 14.4 psi?

absolute pressure = 10 psi + 14.4 psi = 24.4 psi

Gage Pressure. When measuring pressures in many everyday situations, it is the pressure *above* atmospheric pressure that is of interest or importance. It is common practice to call this pressure *gage pressure*, since it is often read on an instrument called a pressure gage. When the gage reads "zero," the pressure at the measurement point is actually equal to the atmospheric pressure at that location. Figure 5-1 shows a graphical representation of atmospheric, gage, and absolute pressures.

Pressure Difference. There are many situations in which the difference in pressure between two points is important. This difference is sometimes referred to as "pressure differential" or "delta p". The latter term also frequently is shown as Δp. Figure 5-2 shows situations where a pressure difference exists (a) across a barrier and (b) between two points in a flow system.

Other Pressure Units. In addition to psi, pressure can also be expressed in pounds per square foot (psf); atmospheres (atm); and, for SI units, newtons per square meter (Pascals or

Figure 5-1 Gage pressure is the pressure above atmosphere pressure. Absolute pressure is the sum of atmospheric pressure and gage pressure.

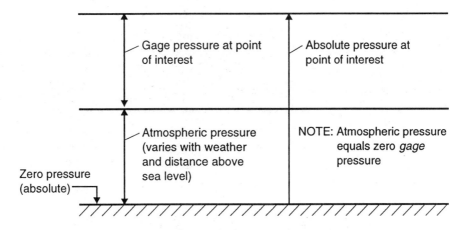

Figure 5-2 The pressure difference ($p_1 - p_2$) represents the net effect of two pressures acting in a system. It can be **(A)** across a barrier or **(B)** between locations in a flow system.

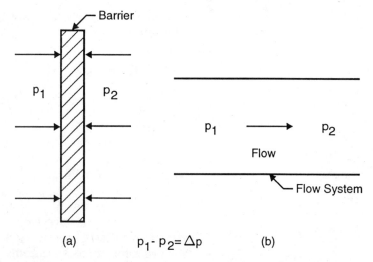

Pa). Because the Pascal is a small unit, kPa (1,000 Pa) is frequently used. The term *atmospheres* here indicates the magnitude of the pressure as compared to the standard atmospheric pressure. A measurement of two atmospheres would describe a pressure equal to twice the standard atmospheric pressure (14.696 psia). Terms related to pressures lower than atmospheric will be considered later in this chapter. Table 5-1 lists various pressure conversion factors.

TABLE 5-1 Pressure Conversion Factors

	Pressure Unit Conversions
1 psi	= 144 psf
	= 6.895 kPa
	= 27.7" water
	= 2.04" mercury
	= 51.7 mm mercury
1 psf	= 0.00694 psi
	= 47.88 Pa
1" water	= 0.036 psi
	= 5.20 psf
	= 0.249 kPa
	= 0.0735" mercury
1' water	= 62.4 psf
	= 0.433 psi
1 atmosphere	= 14.7 psi
	= 1.013 bar
1 torr	= 0.1333 kPa
	= 1.00 mm mercury

EXAMPLE 5-3

What pressure in Pascals corresponds to 7 psi?

$$p = 7 \text{ psi} \left(\frac{6{,}895 \text{ Pa}}{\text{psi}}\right) = 48{,}265 \text{ Pa}$$

Formulas for Calculating Pressure

Hydrostatic Pressure. In a liquid as shown in Figure 5-3, the pressure increases as h, the distance from the fluid surfaces, increases. This increase is due to the weight of the fluid above the point in question. The formula for calculating the pressure at any point is

$$p = \gamma h \qquad (5\text{-}1)$$

where p is the pressure in pounds per square foot, γ the *specific weight* of the fluid in pounds per cubic foot, and h the distance from the surface in feet. The use of these units assures that the units on both sides of Equation 5-1 will be the same.

Figure 5-3 Pressure increases with depth in a fluid.

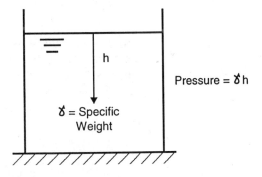

EXAMPLE 5-4

What is the pressure in psi at a location 6 inches below a water surface? Use a specific weight for water of 62.4 lbs/ft³.

$$p = (62.4 \text{ lbs/ft}^3)(0.5 \text{ ft})(\text{ft}^2/144 \text{ in}^2) = 0.217 \text{ psi}$$

Note that the pressure at the fluid surface in Figure 5-3 is atmospheric or, nominally, 14.7 psi. The 0.217 psi calculated in Example 5-4 is a "gage" pressure, or the pressure above atmospheric pressure. Thus, the pressure at the depth of one-half foot could be expressed as 0.217 psi gage (sometimes written psig) or 14.7 + 0.217 = 14.917 psi absolute (sometimes written psia). It is common practice when referring to "gage" pressures to omit the word "gage" and the "g" in "psig." In any particular instance, if there is a chance that confusion could occur, psig, psia, gage pressure, or absolute pressure should be specified.

Also from Figure 5-3, if the fluid were mercury, the pressure would be 13.6 times as large because mercury's specific weight is 848.6 pounds per cubic foot. The ratio of the specific weight of any material to the specific weight of water is called its *specific gravity*, or SG. The specific gravity for mercury is thus 13.6.

EXAMPLE 5-5

If the specific weight for steel is 490 lbs/ft³, what is its specific gravity?

$$SG = \frac{490}{62.4} = 7.85$$

The term *head* is sometimes used to indicate pressure. Used in this manner, it describes a pressure in terms of a column of a particular fluid. Thus, if a pressure were given as "a head of one foot of water," the pressure would be 62.4 lbs/ft² or 0.433 psi. Note that this number can be obtained from Equation 5-1 with $\gamma = 62.4$ lbs/ft³ and h = 1 foot.

EXAMPLE 5-6

Someone tells you that the pressure at a particular location in a body of water is 40 feet of head. What is the pressure at that location?

$$p = (62.4 \text{ lbs/ft}^3)(40 \text{ ft})(\text{ft}^2/144 \text{ in}^2) = 17.33 \text{ psi}$$

The pressures discussed thus far have been for fluids at rest or *static* pressures. The pressure at A in Figure 5-4 is considered a static pressure even though fluid is flowing in the pipe. The probe at B is measuring the effect of stopping

Figure 5-4 Impact pressure includes the effect of stopping the fluid in motion.

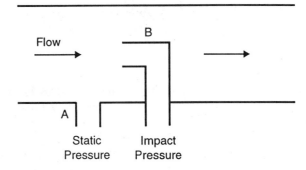

the fluid at its entrance and thus consists of the *dynamic pressure* added to the static pressure. The sum of these two pressures is sometimes called the *total* or *impact* pressure. More will be said about these pressures in Chapter 9.

Note from Figure 5-3 and Equation 5-1 that the pressure below the surface does not vary from one point to another unless one of the points is higher or lower in the fluid. Along any horizontal line, the pressure does not vary.

The fact that the pressure varies with depth and does not vary at any given depth leads to what is sometimes called the *"hydrostatic paradox."* In Figure 5-5, the liquid surfaces will be at the same height (all surfaces are at atmospheric pressure). As the depth increases below each surface, the pressure increases by the same amount. Thus, the pressure at the base of the container under all surfaces is the same. (It is the forces on the walls enclosing the fluid that vary as fluid depth increases.)

Vacuum. Pressures less than atmospheric pressure are usually referred to as either "negative" gage or "vacuum." Both are usually indicated as an amount below atmospheric pressure. For example, if the atmospheric pressure is 14.7 psi, the greatest negative pressure or vacuum possible at that location is 14.7 psi. This would correspond to what is sometimes called a perfect vacuum. Also, "–2 psig" would correspond to

Figure 5-5 The pressure at the base of each container is the same. This is known as the hydrostatic paradox.

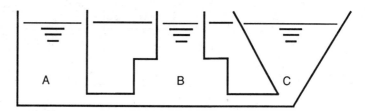

two psi below atmospheric pressure. Values are sometimes given in terms of a column of mercury or water. Thus, a vacuum of 0.5 inch of water would be a pressure less than atmospheric by an amount equal to a 0.5 inch water column.

Vacuum can also be expressed above absolute zero pressure in terms of *torr*. One torr is the equivalent of one millimeter of mercury. Figure 5-6 shows a graphical representation of atmospheric, negative gage, and absolute pressures.

The different methods of measuring and expressing vacuum can be confusing. The following may be helpful:

Consider 7 psia, that is, seven psi above absolute zero pressure.

a. If the atmospheric pressure is 14.7 psia, the pressure could be expressed as 14.7 − 7 or 7.7 psi below atmospheric; it could also be given as a vacuum of 7.7 psi or as −7.7 psig.

b. Because one psi is equivalent to 51.7 millimeters of mercury, the pressure could be given as (7)(51.7) = 361.9 mm Hg absolute or as 361.9 torr.

c. Note that it could also be given as (7.7)(51.7) = 398.1 mm Hg vacuum.

EXAMPLE 5-7

An absolute pressure gage reads 5.0 psia. What is the vacuum at that location if the atmospheric pressure is 14.6 psi?

$$\text{vacuum} = 14.6 - 5.0 = 9.6 \text{ psi}$$

Buoyancy. The term *buoyancy* is used to denote the upward force on an object that is immersed or floating in a fluid. This

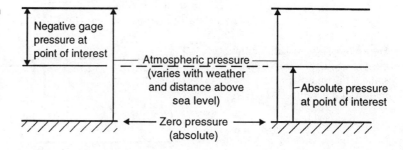

Figure 5-6 Vacuum can be given as a negative gage pressure or as an absolute pressure.

force is the result of the difference in pressure on the various surfaces of the object. The usual formula for calculating the buoyant force is

$$B = \gamma V \tag{5-2}$$

where B is the buoyant force in pounds, γ the specific weight of the fluid in pounds per cubic foot, and V the volume of the fluid displaced by the object. From Figure 5-7, if the object has a volume of three cubic feet and the fluid is water (γ = 62.4 lbs/ft³), the buoyant force would be (62.4)(3) = 187.2 pounds if the object were completely submersed as in Figure 5-7a. If half of the volume of the object were submersed as in Figure 5-7b, the buoyant force would be (62.4)(1.5)= 93.6 pounds. As the object is moved toward the bottom of the container, as in Figure 5-7c, the buoyant force remains 187.2 pounds *as long as fluid pressure can act on the bottom surface*. Note from Figure 5-7d that if the bottom of the container has a "step" in it, there is no buoyancy due to "displaced" fluid because no fluid pressure can get "under" the step. Remember that buoyancy is due to pressure differences and if there is no pressure "underneath" an object or part of an object, there is no buoyancy on that object or part of an object. From Figure 5-8, there is a buoyant force on the object due to the projections, but none due to the center part of the body.

EXAMPLE 5-8

A cubical block of material two feet on each side floats in water with six inches of the block above the water surface. What is the buoyant force on the block?

$$B = (62.4 \text{ lbs/ft}^3)(2 \text{ ft})(2 \text{ ft})(1.5 \text{ ft}) = 374 \text{ lbs}$$

Figure 5-7 The buoyant force on the object in **(A)** and **(C)** is the same. Since only half of the object is immersed in **(B)**, the buoyant force is reduced by half. No pressure gets "under" the object in **(D)**; therefore, there is no buoyant force.

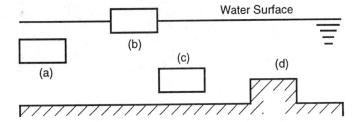

Figure 5-8 If the water pressure does not get beneath the object, the buoyant force is due only to the pressure difference above and below the parts of the object extending to the sides.

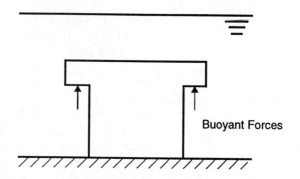
Buoyant Forces

Pascal's Law. Another important phenomenon relating to fluid pressure is frequently referred to as *Pascal's Law*. It states that pressure applied to an enclosed fluid is transmitted to every part of the fluid and to the walls of the vessel containing the fluid. As an example, consider the object shown in Figure 5-9 which demonstrates what is sometimes called the principle of the hydraulic press. The force, F_L, exerted on the piston on the left results in pressure

$$p = \frac{F_L}{A_L} \tag{5-3}$$

where A_L is the cross-sectional area of the piston. Since from Pascal's Law the pressure is the same in both cylinders, the force on the right piston can be obtained from

$$p = \frac{F_R}{A_R} \tag{5-4}$$

where A_R is the cross-sectional area of the right piston (in this type of device, the variation of pressure with vertical distance is usually small compared to the applied pressure). Thus,

$$\frac{F_R}{A_R} = \frac{F_L}{A_L} \tag{5-5}$$

or

$$F_R = \frac{A_R}{A_L} F_L \tag{5-6}$$

From Equation 5-6, it can be seen that the force F_L is "magnified" by an amount equal to the ratio of the piston areas. The hydraulic press principle is used in such applications as car hoists, barber chairs, and the like.

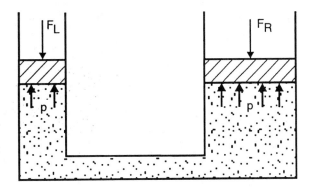

Figure 5-9 The hydraulic press is a force-multiplying device.

EXAMPLE 5-9

Consider a horizontal piston arrangement with an area of 0.5 ft² on the left piston and 1.5 ft² on the right. If the force on the left piston is 100 lbs, what is the force delivered by the right piston? What is the pressure in the system?

$$\text{force on right piston} = \frac{1.5}{0.5} \times 100 = 300 \text{ lbs}$$

$$p = \frac{300 \text{ lbs}}{1.5 \text{ ft}^2} \left[\frac{\text{ft}^2}{144 \text{ in}^2} \right] = 1.389 \text{ psi from right piston}$$

$$p = \frac{100 \text{ lbs}}{0.5 \text{ ft}^2} \left[\frac{\text{ft}^2}{144 \text{ in}^2} \right] = 1.389 \text{ psi from left piston}$$

MEASURING DEVICES

Manometers

Manometers are not as widely used as they once were, but they are used occasionally and are a good illustration of one of the principles of pressure measurement.

106 / PRESSURE

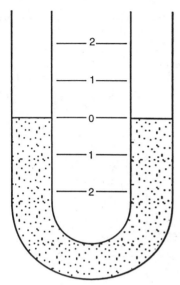

Figure 5-10 A simple U-tube manometer.

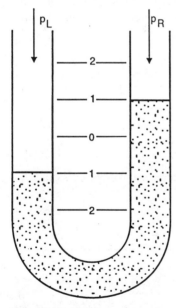

Figure 5-11 The pressure on the left side of the U-tube manometer, p_L, is higher than on the right, p_R.

U-Tube Manometer. A *U-tube* manometer consists of a glass tube shaped like the letter U with a scale marked off in appropriate units as shown in Figure 5-10. The zero mark appears at the center; and when pressure is introduced on the left, as shown in Figure 5-11, the level of the dark fluid moves down, which, in turn, causes the level of the dark fluid in the right column to move up. The vertical distance between the two levels is a measurement of a fluid column (see Equation 5-1) and is used in determining the pressure of interest.

In reading a manometer, care should be taken to read the liquid surface or meniscus at its center. For water, this is the bottom of the surface; for mercury, it is the top of the surface.

One application of a U-tube manometer is shown in Figure 5-12. Fluids flow in Pipes A and B in the direction into or out of the page. The manometer is set up to measure the pressure difference between the center of Pipe A and the center of Pipe B. Using Equation 5-1 and starting at the center of Pipe A:

$$p_A + \gamma_A(L + h) - \gamma_m h - \gamma_B(L + H) = p_B \qquad (5\text{-}7)$$

where p_A and p_B are the pressures at the centers of Pipes A and B and γ_A, γ_B, and γ_m are specific weights of the flowing fluids and the manometric fluid, respectively. A key to understanding manometer calculations is that the pressure in the manometer fluid at the level of line c–c is the same in both legs of the manometer. As an aid in understanding this, consider that the pressure in the manometer fluid along line e–e is constant (there is no vertical change between any two points on the line). Moving up from line e–e in either of the legs, the pressure changes by an amount related to the vertical distance above e–e. Thus, at c–c, the pressure in both legs is the same.

Rearranging Equation 5-7 to form a pressure difference:

$$p_A - p_B = -\gamma_A L - \gamma_A h + \gamma_m h + \gamma_B L + \gamma_B H \qquad (5\text{-}8)$$

For further interpretation, Equation 5-8 can be written as:

$$p_A - p_B = \gamma_B H + (\gamma_B - \gamma_A)L + (\gamma_m - \gamma_A)h \qquad (5\text{-}9)$$

Consider the following special cases:

1. The centers of Pipes A and B are at the same elevation (i.e., H = 0). This eliminates the first term on the right side of Equation 5-9.
2. The same fluid is flowing in Pipes A and B (or both fluids have the same specific gravity); that is, $\gamma_A = \gamma_B$. This eliminates the second term on the right side of the equation.
3. The specific weight of fluid A is much smaller than that of the manometer fluid (as could be the case if fluid A were a gas and the manometer fluid a liquid). This eliminates γ_A from the right side of the equation.

If all three special cases were in effect:

$$p_A - p_B = \gamma_m h \tag{5-10}$$

which is similar in form to Equation 5-1 and shows that the differential pressure in this special case is equivalent to the pressure under a column of manometer fluid h units high. In many applications, Equation 5-10 is sufficient for expressing the pressure difference of interest even though the conditions of the special cases are not totally met. Note that the diameter of the manometer tube (and thus its cross-sectional area) does not enter into the calculation.

EXAMPLE 5-10

Using Figure 5-12, calculate the pressure difference $p_A - p_B$ for the following conditions:

$$\gamma_A = 62.4 \text{ lbs/ft}^3; \gamma_B = 56.2 \text{ lbs/ft}^3; \gamma_m = 849 \text{ lbs/ft}^3;$$

$$H = 2 \text{ ft}; L = 1 \text{ ft}; h = 2.5 \text{ ft}.$$

$$p_A - p_B = 56.2 (2) + (56.2 - 62.4)(1) + (849 - 62.4) 2.5$$

$$= 112.4 - 6.20 + 1,966 = 2,072 \text{ lbs/ft}^2$$

U-Tube Manometer with Open Leg. For some manometer applications, the pressure of interest is that in Pipe A; and the other manometer leg is open to atmospheric pressure (in place of connecting to Pipe B). Thus, $p_B = 0$. Consideration of Equation 5-8 and the above development will, in many cases, give the simple result:

$$p_A = \gamma_m h \tag{5-11}$$

Such an application is shown in Figure 5-13.

Figure 5-12 A manometer set-up for measuring the pressure difference between two flow lines.

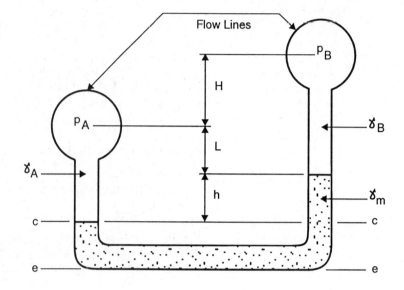

EXAMPLE 5-11

A manometer with one leg open to the atmosphere has an h reading of 9 inches. If the manometer fluid is water, what is the pressure being measured? See Figure 5-13 for set-up.

$$p = (62.4 \text{ lbs/ft}^3)(9 \text{ in})(\text{ft}/12 \text{ in})(\text{ft}^2/144 \text{ in}^2) = 0.325 \text{ psi}$$

Figure 5-13 A manometer set-up for measuring the pressure in a flow line.

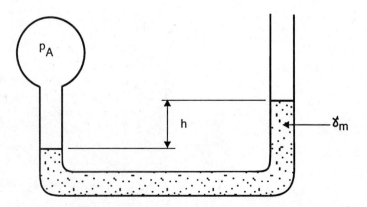

Figure 5-14 An inclined-tube manometer gives more accurate readings because of the longer scale.

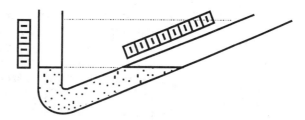

Inclined Manometer. An inclined manometer is shown in Figure 5-14. Usually used to measure rather low pressures, the low-pressure arm is set at an angle so that the scale to be read is longer. This leads to a higher-percentage accuracy. Because inclined manometers can be used to measure the draft on a furnace, they are sometimes called draft gages.

Well Manometer. Another manometer for measuring low pressures is shown in Figure 5-15. In the well manometer, the cross-sectional area of one leg is much larger than the other. Because of this, there is very little vertical movement of the larger-area column and the pressure of interest is essentially indicated by the height of the smaller-area column.

To illustrate: If a pressure were applied to a well leg with a cross-sectional area of 100 in^2, the fluid level movement in the smaller leg (with a cross-sectional area of 0.5 in^2) would be 200 times that in the well. For many applications then, the movement in the well can be ignored and readings simply taken from the smaller-area leg. In applications where it is desired to account for movement in the well, the scale markings on the smaller leg can be specially placed based on the ratio of the cross-sectional areas.

Figure 5-15 A well-type manometer.

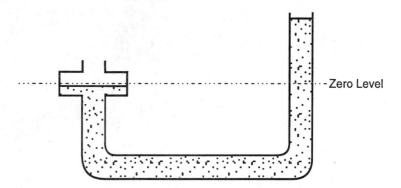

Water is a common manometer fluid. In the past, mercury was often used; but because of potential health hazards, its use has been on the decline. It is possible to obtain manometer fluids with specific weights varying from less than that for water to around three times greater than water.

Gages

One of the classifications of devices used to measure pressure is simply called *gages*. A typical gage is shown in Figure 5-16. There are several mechanical elements that can be used as the pressure sensors, including diaphragms, bellows, capsules, and tubes (see Figure 5-17). Pressure causes these sensors to deform, and this deformation is converted to the movement of an indicating pointer.

Bourdon Tube. Perhaps the most frequently used of these devices is the Bourdon tube, which was patented in 1840 (see Figure 5-18). As pressure is applied to the inside of a Bourdon tube, the difference in pressure between the inside and out-

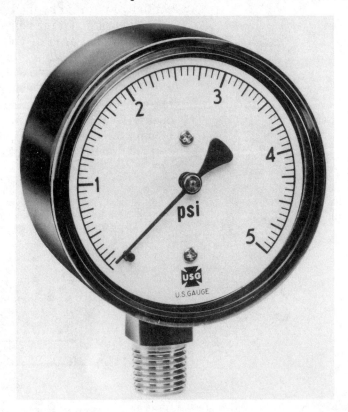

Figure 5-16 A pressure gage for measuring up to 5 psi *(Photo courtesy of Ametek, Inc.)*

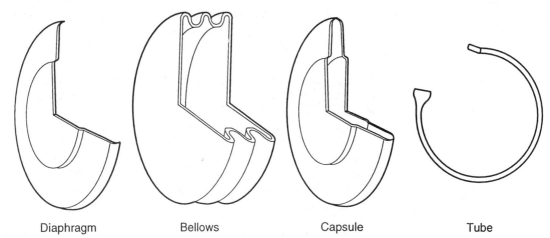

Diaphragm   Bellows   Capsule   Tube

Figure 5-17 Sensing elements for pressure gages.

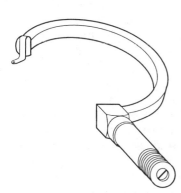

Figure 5-18 Bourdon tube.

side of the tube causes it to straighten slightly. One end is fixed, and the movement of the other end is converted to a pointer movement. Tubes are usually made from seamless metal tubing with wall thicknesses in the range of 0.01 to 0.05 inch. Material used includes beryllium copper, steel, and stainless steel. If a Bourdon tube is slightly overloaded, recalibration may be necessary. Severe overload leads to a ruined gage. Most gages can handle an overload of approximately 30–40% of their rated maximum without permanent damage. Tubes sometimes are formed into helical or spiral shapes for measuring higher pressures. In general, tube-type gages are available for ranges of a few psi up to around 100,000 psi.

Diaphragm. Some diaphragms are made of very thin, treated, nonporous rubber or plastic and require very little pressure to deform (they should be inspected periodically because they may become stiff or develop leaks). A spring is used in combination with the diaphragm to produce the desired reading. Diaphragm gages are very sensitive to overload; thus, care should be taken to avoid this situation. Other diaphragms are made of either flat or corrugated metallic disks. Typical diaphragm pressure ranges are usually in the order of 0–0.5 inch to 0–10 inches of water. Applications include tractors, trucks, and industrial equipment where the small movement is advantageous.

Capsule. Two diaphragms can be joined to form a "capsule." Capsules can be used alone or in stacks, depending on the pressure range desired. The deflection of a capsule depends on its diameter, the material thickness and elasticity, and the design. Materials used include phosphor bronze, stainless steel, and an iron–nickel alloy. Typical capsule pressure ranges go up to 50 psi.

Diaphragm and capsule elements themselves respond to pressure changes very quickly. The length and diameter of tubing used to connect them affect the overall response time. In general, the lower the pressure to be measured, the shorter the length and the smaller the diameter should be for the connecting lines.

Bellows. A bellows is an element with a series of folds or corrugations called convolutions. Materials used include brass, copper–nickel alloys, Monel, steel, and stainless steel. Bellows are usually used in conjunction with a spring to simplify calibration and adjustment. Pressure ranges for bellows can be up to several hundred psi.

Barometers

Instruments used for measuring atmospheric pressure are called *barometers*. A simple barometer is shown in Figure 5-19. A glass tube, sealed at one end, is filled with mercury, inverted, and placed in a pan of mercury. The mercury in the tube then moves downward, leaving a vacuum above it. The atmospheric pressure is indicated by the height of the mercury column in the tube above the level in the pan. Another atmospheric pressure device is called an *aneroid barometer*. It consists of a pressure capsule from which all air has been

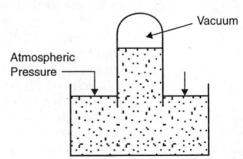

Figure 5-19 A simple barometer.

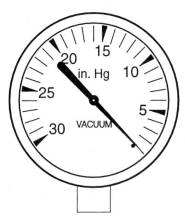

Figure 5-20 A vacuum gage for measuring to 30 inches of mercury vacuum.

removed. An attached pointer indicates movement related to changes in atmospheric pressure. The term *aneroid* means that no fluid is used in making the measurement.

Vacuum Instruments. For relatively weak vacuums (pressures not far below atmospheric), elastic deformation elements, such as Bourdon tubes, bellows, or diaphragms can be used. Gages using these devices would indicate an amount below atmospheric pressure. Figure 5-20 shows a typical vacuum gage.

A vacuum could also be calculated using an absolute pressure measuring device. If the atmospheric pressure were 14.7 psia and the absolute pressure device read 12.0 psia, there would be a vacuum of 14.7 − 12.0 = 2.7 psi. Most conventional absolute pressure transducers can be used down to 20 torr and, with special sensing devices, down to around 1 torr.

For measuring relatively strong vacuums, other types of devices must be used. Two of these are the Pirani gage and a special thermocouple set-up (thermocouples are discussed in Chapter 7). Both of these methods are based on the fact that heat is conducted and radiated from a heated element placed in the low-pressure region. The heat loss is related to the number of gas molecules per unit volume in the low-pressure region surrounding the element which is, in turn, related to the pressure. These two devices can be used down to around 0.01 psia.

An ionization gage can be used for even smaller absolute pressures (stronger vacuums). This gage uses electrons to ionize the gas whose pressure is being measured and then measures the current flowing between two electrodes in the gas. The number of ions per unit volume depends on the gas pressure and so the current also depends on gas pressure.

The McLeod gage is another device for measuring low absolute pressure. Its operation is based on the fact that the pressure and volume of an enclosed gas are inversely related. By compressing a known volume of low-pressure gas to a higher pressure and measuring the resulting pressure and volume, the original pressure can be calculated. Pressures from less than 1 mm Hg up to around 50 mm Hg can be measured with this device. If the gas whose pressure is being measured contains vapor that could condense, the McLeod gage should not be used.

114 / PRESSURE

Figure 5-21 An electrical pressure transducer *(Courtesy of Validyne Engineering Corp.)*

Transducers

The deformation elements discussed previously in the section on gages can be incorporated in an electrical device to form a pressure transducer (see Figure 5-21). These transducers can produce changes in resistance, capacitance, or inductance.

In general, the resistance transducers are comprised of strain gages (a fine wire formed into a grid) which are distorted by the deformation taking place. The distortion changes the strain gage resistance which, in turn, can be related to the pressure. Another strain-type transducer uses an elastic disk on which a sensing grid has been deposited using crystal diffusion. Resistance transducers typically can detect very small movements and thus very low pressures.

Capacitance transducers are usually made of two conductive plates and a dielectric. Increased pressure moves the plates apart which changes the capacitance. The fluid whose pressure is being measured is the dielectric. A schematic diagram of a capacitance transducer is shown in Figure 5-22.

Inductance transducers contain three important parts: a coil, a moveable magnetic core, and a deformation element. The element is attached to the core; and as the pressure changes, the element causes the core to move within the coil. Alternating current passes through the coil; and as the core moves, the inductance of the coil changes. A reluctance transducer is similar to an inductance transducer except that it has a permanent magnet as part of its mechanism. A schematic diagram of an inductance transducer is shown in Figure 5-23.

Other types of transducers include piezoelectric and carbon pile. The piezoelectric effect occurs in quartz crystals.

Figure 5-22 Schematic diagram of a capacitance device for sensing pressure.

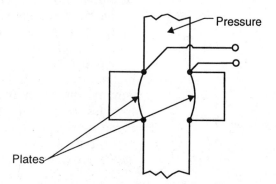

Figure 5-23 Schematic diagram of an inductance device for sensing pressure.

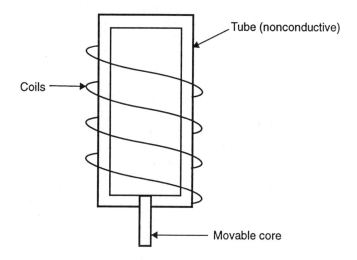

When pressure is applied to a suitably prepared crystal, an electrical voltage across points in the crystal can be measured and related to the pressure. Because of the characteristics of the electrical circuits in which they are placed, piezoelectric transducers are not usually used for measuring static (unchanging or slowly changing) pressures. Some piezoelectric transducers have very high sensitivities and high frequencies which allow them to be used over a wide range of applications. In a carbon pile transducer, pressure is applied to a confined mass of carbon particles. The electrical resistance of the carbon pile can be related to the pressure of interest. Schematic diagrams of carbon pile and piezoelectric transducers are shown in Figures 5-24 and 5-25.

Figure 5-24 Schematic diagram of a carbon pile device for sensing pressure.

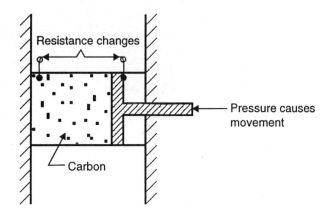

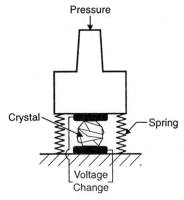

Figure 5-25 Schematic diagram of a piezoelectric device for sensing pressure.

Differential Pressure Devices

Most of the pressure-measuring devices discussed thus far can be used for measuring differential pressure. The application of a manometer for doing so is shown in Figure 5-12.

In a bellows-type differential meter (see Figure 5-26), each of the pressures are fed into a different bellows. Because the pressures are different, one bellows deflects more than the other. The deflection is then used to indicate the pressure difference.

In a bell-type differential meter (see Figure 5-27), a bell is floated in a pool of liquid in the meter body. The high pressure is introduced to the underside of the bell and the low pressure to the outside. Movement of the bell is opposed by a spring whose deflection can be related to the pressure difference.

Another type of differential pressure device used in the past is the weight–balance-type differential meter. An example is the ring balance in which mercury contained in the ring is displaced by the differential pressure. The ring is balanced

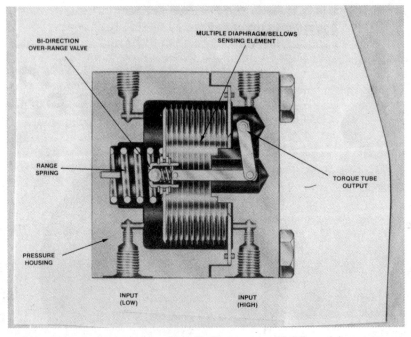

Figure 5-26 Typical cross-section of bellows-operated differential pressure gage *(Courtesy of Mid-West Instruments)*

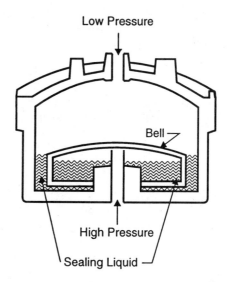

Figure 5-27 Bell-type differential pressure meter.

on a sharp edge and counterbalanced by a weight. The amount of weight needed for balancing the ring is then related to the pressure differential.

APPLICATION CONSIDERATIONS

Obtaining meaningful pressure information requires consideration of several items.

Selection

The selection of a pressure-sensing device for a particular situation involves factors such as pressure range, accuracy required, and response time.

Table 5-2 lists typical approximate ranges for several devices. In selecting a device, its range should match that of the expected application as closely as possible in order to obtain the best accuracy. The reason for this is that the accuracy of most devices is given as plus or minus a certain amount of the maximum device reading. Thus, if a device had a recommended maximum reading of 100 psi with an accuracy of ±1%, it could be in error by as much as 2% at 50 psi, and 4% at 25 psi. Pressure-sensing devices are available with varying accuracies. In general, the greater the accuracy, the higher the cost. Typical Bourdon tube accuracies might be anywhere from ±0.25% to ±1% of the maximum reading. Most pressure-

Table 5-2 Approximate Pressure Ranges for Pressure-Sensing Devices

	Minimum Range	Maximum Range
Diaphragm	0–0.5" water	400 psi
Bellows	0–5" water	800 psi
Capsule	0–0.5" water	50 psi
Bourdon tube	0–12 psi	100,000 psi
Spiral bourdon	0–15 psi	40,000 psi
Helical bourdon	0–50 psi	80,000 psi
Bell (differential)	0–1" water	40" water
Ring manometer	0–20" water	400" water
Strain gage instrument	—	+100,000 psi

sensing devices respond quickly to changes in pressure. If response time is important in a particular application, information from the supplier should be obtained to ensure satisfactory results.

The term *span* for a pressure-measuring device refers to the difference between the maximum and minimum readings for the device. Usually when applied to gages, the span is the same as the maximum reading on the gage. Thus, a gage with a scale of 0 to 100 psi would have a 100 psi span. In some special situations, the lowest reading on the gage might not be zero. A gage with scale of 20 to 100 psi would have a span of 80 psi.

Specifications for pressure transducers may include information regarding hysteresis (which can be thought of as the sensing device's ability to "forget" immediately proceeding pressures), error, and linearity (constant ratio between output and input). In addition, differential devices may have a *proof* pressure which indicates the pressure that the transducer itself can tolerate. *Overpressure* is referenced to the measurement range and is usually given as one of the following: the amount of excess pressure that will not cause a shift in the sensing device's "zero"; the amount that will cause such a shift; or the amount that will cause permanent damage.

Installation

With regard to the installation of pressure-sensing devices, the following should be kept in mind:

1. The distance between the source of the pressure being measured and the sensing device should be kept as short as possible to keep response time to a minimum.

2. Air or gas (other than the fluid whose pressure is being measured) trapped in the lines connecting the pressure source and the sensing device introduces errors in liquid pressure measurement. Also, similar problems can occur with condensed liquids in gas pressure measurement. If possible, connecting lines should rise from source to device in gas systems and fall from source to device in liquid systems.
3. When a pressure-sensing device is located below a pipe and liquid pressure measurement is being made, the pressure due to the column of fluid in the connecting line will introduce an error into the reading. Calculations should be made to decide whether the reading should be corrected. In some cases, the zero adjustment of the transducer can be used to make this correction.
4. Connecting lines should have the proper valves and fittings so that pressure-sensing devices can be removed or replaced easily.
5. Pulsating pressures in the connecting lines can cause difficulty in reading the sensing instrument and, in more severe cases, cause excessive wear and loss of calibration. These pulsations can be reduced by adding resistance and capacity to the connecting line. One of the simplest methods for adding resistance involves a needle valve installed in the connecting line. Capacity can be added with a small-diameter pipe connected in parallel with the existing connecting line.
6. When the pressure of interest is from a corrosive fluid, the fluid must be kept from the sensing device (unless the device itself is made from a noncorrosive material). There are two ways this can be accomplished: (a) using an inert liquid in the sensing device and adjoining lines and (b) using a sensing device with a completely sealed system.
7. Overrange protection for the sensing device can be included in the systems. It is often used for low-range devices sensing air or gas pressures.

Calibration

Pressure-sensing devices are usually calibrated before they leave the factory. Occasionally, there is need for recalibrating. This can be done in several ways. A liquid manometer can be used for low-range devices. Devices with higher ranges can weight tester. In the standard test gage method, the device under consideration is hooked up in a common manifold with the standard gage (a very accurate gage used only for calibration purposes). Using a screw-controlled plunger, the desired pressure can be applied, and the reading of the device being calibrated can be compared to the standard gage.

Use of a dead-weight tester eliminates the possibility that the standard gage can be in error. In this method, a series of weights is supported by a piston which is positioned by the pressure in the system. Because of the design of the dead-weight tester, small loads can create very high pressure. With proper weights and careful handling, accuracies with the dead-weight tester are very good. Figure 5-28 shows a dead-weight tester.

Figure 5-28 A dead-weight tester for calibrating pressure devices *(Courtesy of DH Instruments, Inc., Tempe, Ariz.)*

Protection

In some applications, another fluid is used in addition to the manometer fluid. Its purpose is to keep the fluid whose pressure is being measured from contacting the manometer fluid. The reasons for using this additional or "sealing" fluid include protection of the manometer from the fluid whose pressure is being measured (due to corrosiveness for instance as mentioned in the section on installation) or to protect the manometer itself. It may be necessary to correct for the density difference between the sealing fluid and the other fluids. If so, an approach similar to that applied in Figure 5-12 can be used. Figure 5-29 shows an application of a sealing fluid. Using an approach similar to that in Equation 5-7:

$$p_A + \gamma_A(L + L_1) + \gamma_s (L_2 + h) \\ - \gamma_m h - \gamma_s(L_1 + L_2) - \gamma_B (L + H) = p_B \quad (5\text{-}12)$$

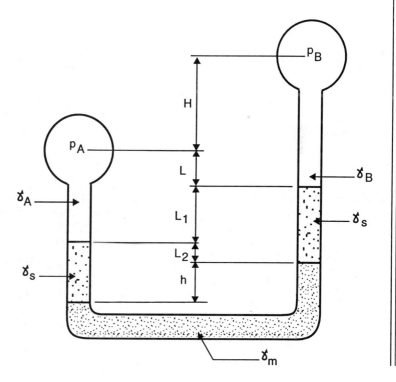

Figure 5-29 A manometer set-up with sealant fluid in each leg.

Rearranging gives

$$p_A - p_B = (\gamma_m - \gamma_s)h + (\gamma_s - \gamma_A)L_1 \\ + (\gamma_B - \gamma_A)L + \gamma_B H \quad (5\text{-}13)$$

Following the reasoning used after Equation 5-7, if the pipes are at the same elevation, the last term in Equation 5-13 is zero and the pressure difference $p_A - p_B$ depends on three specific weight differences and three column lengths. If the fluids in Pipes A and B have the same specific weight, the term $(\gamma_B - \gamma_A)L$ is zero. If the specific weight of the sealing fluid is close to that of the fluid flowing in the pipe and the term $(\gamma_s - \gamma_A)L$ is small, the resulting equation is then

$$p_A - p_B = (\gamma_m - \gamma_s)h \quad (5\text{-}14)$$

Just as for Equation 5-10, Equation 5-14 is sufficiently accurate in many cases for making corrections due to the use of a sealing fluid.

REVIEW MATERIALS

Important Terms

pressure
atmospheric pressure
absolute pressure
gage pressure
Pascal
specific weight
hydrostatic pressure
specific gravity
head
dynamic pressure
static pressure
total or impact pressure
hydrostatic paradox
vacuum
torr
buoyancy
Pascal's Law
U-tube manometer

meniscus
inclined manometer
well manometer
gage
Bourdon tube
capsule
diaphragm
bellows
barometer
aneroid barometer
piezoelectric effect
differential pressure
span
proof pressure
overpressure
dead-weight tester
sealing fluid

Questions

1. Why are pressure measurements important in manufacturing instrumentation?
2. Name some metals used for elastic elements in pressure gages. What are some metal properties that would be important?
3. What is a possible effect of pulsing pressure on the performance of a pressure gage?
4. Why are liquid pressure systems filled from their lowest point?
5. What type of differential pressure device would be used for following rapid pressure changes?
6. What would be desirable properties for liquids to be used to seal pressure gage lines?
7. What sensing device could be used for measuring a pressure that is usually between 400 and 600 psi and could go up to 700 psi?
8. How does a dead-weight tester work, and what precautions should be used when using one?
9. Why might high-pressure gages not be used for making differential pressure measurements?
10. What happens to the response rate of a gas pressure gage when a long lead is required to reach the gage?
11. Why must the line to a pressure gage be filled with liquid when the pressure of a liquid is being measured?
12. Why is an obstruction to flow sometimes purposely put in the line to a pressure gage?
13. What is the difference between force and pressure?
14. Pressure gages are located at different points along a pipeline. Is the pressure difference between the readings an absolute or a gage pressure?
15. Absolute pressure is equal to gage pressure subtracted from atmospheric pressure. True or False? Explain.
16. One of the reasons for making absolute pressure measurements instead of gage pressure measurements is that the effect of atmospheric pressure variation is eliminated. True or False? Explain.
17. Always read the bottom of the meniscus in a manometer. True or False? Explain.
18. What type of pressure-measuring instrument is a draft gage?
19. How does a bell pressure gage work?
20. There are two faucets in a water line. One is open with flow coming out. Why does the flow decrease if the other faucet is opened?

Problems

1. What is the pressure in psig at the bottom of a tank which is filled to a depth of 25 feet with fresh water?
2. If a skin diver goes to a depth of 40 feet in fresh water, what is the pressure on his body?

124 / PRESSURE

3. If the water supply for a city were 200 feet above the lowest point in the city, what would be the maximum available water pressure?
4. The expansion tank in a hot-water home heating system is open to the atmosphere and 28 feet above a gage on the basement furnace. What will the gage read in psi?
5. Some common pressures are automobile tire = 28 psi; water pressure = 60 psi; football = 13 psi. What are the equivalent pressures in kPa?
6. If a Bourdon gage reads 25 psi, what is the absolute pressure in Pa?
7. Convert the following pressures into inches of water: (a) 5 psi and (b) 0.6 psi.
8. Convert the following pressures into inches of mercury: (a) 50 psi and (b) 14.0 psi.
9. How many psi are equal to 432 psf? How many psf are equal to 1.5 psi?
10. A tank of oil with specific gravity 0.8 is to be emptied through a siphon. How much pressure must be introduced into the tank to start flow if the siphon is 5 feet above the fluid level as indicated in Figure 5-30?

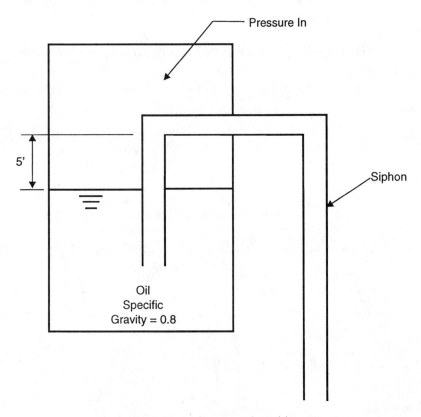

Figure 5-30 Figure for use with Problem 10.

11. A manometer is used for measuring the pressure in a tank as shown in Figure 5-31. If the atmospheric pressure is 29.7 inches of mercury, what is the pressure in the tank in psia? Consider the contents of the tank to be a gas.

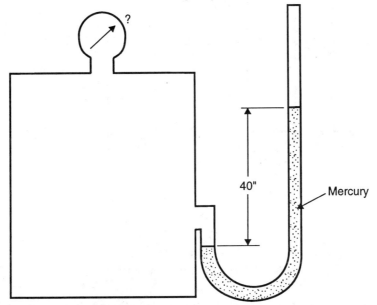

Figure 5-31 Figure for use with Problem 11.

12. A mercury barometer has no air space above the mercury in the closed tube. If the mercury column is 26 inches high, what is the atmospheric pressure?
13. The left leg of the U-tube manometer has a water column 40 inches high and is in equilibrium with a 20 inch column of liquid A in the right column. What is the specific weight of liquid A in Figure 5-32?
14. A tank is partially filled with water and oil. If the specific weight of the oil is 51 lbs/ft³ and the thickness of the oil and water layers are each 4 feet, what is the pressure at the bottom of the tank?
15. If the atmospheric pressure is 29.5 inches of mercury and a vacuum gage reads 7 inches of mercury, what is the absolute pressure in psi?
16. A standard atmosphere is 101.325 kPa. What would be the equivalent psf, meters of water, inches of an oil with specific gravity 0.90, and torr?
17. Assume that an airplane door is 7 feet high and 4 feet wide. If the pressure inside the plane is 14 psia, what is the force on the door when the plane flies at an altitude where the pressure is 2 psia?
18. If the gage pressure is 24 psi at a certain level, what is the absolute pressure? If the gage pressure is −5 psi, what is the absolute pressure?

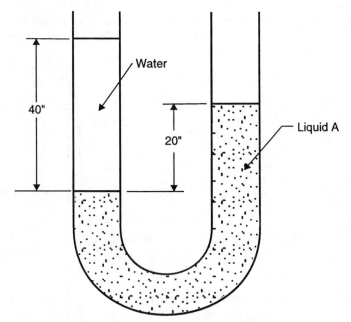

Figure 5-32 Figure for use with Problem 13.

19. Assume a submarine sinks to a depth of 300 feet. What is the absolute pressure on the sub? Use 1.025 as the sea water specific gravity.
20. A Bourdon gage has a scale of 0 psi to 200 psi and has an accuracy of ±1% of span. If the gage reads 150 psi, what will be the limits for the correct pressure?
21. A pressure gage can measure from a low of 200 psi to a high of 700 psi. What is the span of the gage?
22. The pressure on the legs of a mercury manometer are 100 psi and 80 psi. What is the difference in height of the mercury columns?
23. If a column of manometer fluid is 36 inches high with a specific weight of 0.328 lb/in^3, what is the pressure of interest? Assume fluid of interest is a gas with a much lower specific weight.
24. A U-tube manometer (Figure 5-33) uses mercury as the measuring fluid and water as the transmitting fluid in both legs. What would the vertical distance between the mercury levels be if the applied differential pressure is 23.2 psi?
25. An open tank is filled with water to a depth of 10 feet. What is the force acting on each square foot of the tank bottom?
26. If a force of 1,000 newtons is applied to a piston with an 8 cm diameter, what is the pressure at the face of the piston in kPa?
27. The diameters of the pistons in a hydraulic press are 6 inches and 60 inches. If the force applied to the smaller one is 200 pounds, what will be the pressure at the larger piston and what force will be available there?

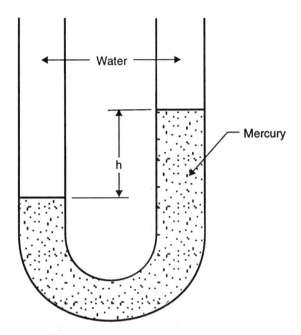

Figure 5-33 Figure for use with Problem 24.

28. The three containers shown in Figure 5-34 have circular bases. What is the pressure and force acting on the base of each if the fluid is water?
29. A stone weighs 240 lbs in air. When submerged in water, it "weighs" only 140 lbs. What are the volume and specific weight of the stone?
30. A piece of wood one foot square and one inch thick has a specific gravity of 0.5. What volume of lead would have to be attached to the bottom of the wood so that the top of the wood would be even with the water surface in a water tank? Specific gravity of lead = 11.3.

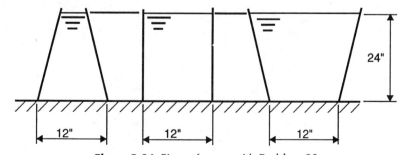

Figure 5-34 Figure for use with Problem 28.

6

Signal Transmission

CHAPTER GOALS

After completing study of this chapter, the student should be able to do the following:

List the principal modes of signal transmission for instrumentation and process control.

Know the most common standardized ranges of values of transmission variables.

Understand the slight differences between converters, transducers, and transmitters.

Discuss why some modes of signal transmission are more suitable for long distances than others.

Design a simple P/I or P/E transducer.

Explain the purpose of a transfer function.

List some of the purposes of signal conditioning.

Understand he simpler concepts of digital transmission.

Discuss the purpose of the span and offset adjustments available on transducers.

The measurement made by an instrument is, in many cases, useless unless it is sent elsewhere for either display or processing. (The exception to this statement is the case of instruments that are self-contained, such as a hand-held voltmeter.) For example, the temperature sensing done by the sensor in a heat-exchange process control unit must be relayed to the controller. The resulting correction signal, known as the manipulated variable signal, must be sent from the controller to the actuator. This sending or relaying of signals is known as transmission. *Transmission* is the transferring of information signals from one point to another. In the case of instrumentation, these signals are the values of variables.

For process control, the transmission of the measured variable signal from the sensor/transducer to the controller and the transmission of the manipulated variable signal from the controller to the actuator are as important as any other phase of the process control operation. If the signal is not transmitted in true, accurate form, then erroneous information is received by either the controller or the actuator and process control will fail. (Accurate transmission of the information signal from a sensor to simply a panel display is also important.)

There are several different ways signals can be sent. The method used will depend on the sensor and controller being used, the distance between the elements of the process control loop, and the degree of precision with which the process needs to be controlled. These different methods of transmission and their advantages and disadvantages are discussed in this chapter.

ELECTRICAL

The relaying of information in the form of an electrical signal is the most common of all modes of transmission. There are three common electrical transmission methods that are used in the industrial instrumentation field: analog, digital, and telemetry. Analog uses a continuously varying amplitude signal, and digital and telemetry use nonamplitude-related signals, often in pulse form. Because pulse-related signals can relay information with less chance of degradation of the signal, the pulse methods are very commonly used for long distances, say, any length greater than approximately a quarter of a mile. Most instrumentation and process control distances are less than that, so the analog electrical form is the most frequently used method. Digital is rapidly gaining in popularity, however, even for the shorter distances.

Analog

Standardized Ranges. Nearly all analog electrical signal systems in the world of instrumentation will be transmitting information using either *current* or *voltage*. This means that it is either the amount of current being sent that indicates the value of the variable of interest, or it is the amount of voltage being sensed by the receiving device that indicates the value of the variable. If current signals are used, they are usually required to fit within one of the two following standardized ranges: 4 to

20 mA or 10 to 50 mA. The range of 10 to 50 mA is one that was in common use in earlier years, but its use is rapidly fading.

If voltage signals are used, they will often fall within two common voltage ranges: 0 to 5 v or 0 to 10 v. Of these two voltage ranges, the range of 0 to 5 v is the more standardized. For example, if one thumbs through an industrial instrumentation catalog, it will be found that many instruments with built-in transducers have outputs of 0 to 12 v, or some similar value, rather than 0 to 10 v. The range of 0 to 5 v is seldom modified, however. It will also be found that of all of the ranges mentioned in this section, the current range of 4 to 20 mA is easily the most common.

The purpose of standardization is to make separate components of instrumentation and process control equipment compatible. In particular, it makes the replacement of parts easier than before.

Transducers. The origin of a measured variable signal is at the sensor. However, as mentioned in Chapter 1, the signal usually needs to be converted to some other energy form in order to be transmitted. For example, there is a temperature-sensing device known as a thermocouple that self-generates a very small voltage, the amount of which is dependent on the temperature sensed by the thermocouple. This voltage is typically in the microvolts range and is too small to be dependably transmitted. To compensate for this, the thermocouple is connected directly to a transducer that amplifies the signal and, in most cases, converts it from a voltage signal to a current signal. It is the conversion of the signal from voltage form to current form that somewhat fits the common definition of the transducer. Figure 6-1 shows an op-amp circuit that performs voltage-to-current (usually abbreviated as "E/I") conversion.

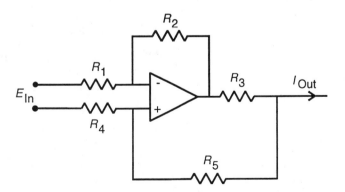

Figure 6-1 Schematic of an E/I converter.

EXAMPLE 6-1

Suppose a thermocouple's maximum output within the temperature range to be measured by the control system was 800 microvolts. What amplification would the transducer need to ensure a maximum output of 10 volts?

$$E_{out} = A \cdot E_{in}$$

where A represents the amplification factor.

$$A = \frac{E_{out}}{E_{in}}$$

$$= \frac{10 \text{ v}}{800 \text{ µv}}$$

$$= 0.0125 \cdot 10^6, \text{ or } 12,500$$

A *transfer function* is an equation that describes the relationship between the input and output of an element. In Figure 6-1, the output current is related to the input voltage by the following transfer function:

$$I_{out} = -\frac{R_2 \times E_{in}}{R_1 R_3} \tag{6-1}$$

The three resistors in Equation 6-1 must be related to the other two resistors by

$$R_1(R_3 + R_5) = R_2 R_4 \tag{6-2}$$

EXAMPLE 6-2

What ratio of R_2 to the product of R_1 and R_3 would be needed to convert a 5 v input signal to a 20 mA output signal? (Ignore the current direction sign.)

$$\frac{R_2}{R_1 R_3} = \frac{I_{out}}{E_{in}}$$

$$= \frac{20 \text{ mA}}{5 \text{ v}}$$

$$= 0.004 \; \Omega^{-1}$$

The negative sign in the equation indicates that the current actually flows in the opposite direction from that indicated by the arrow. The current can be made to flow in the other direction by simply reversing the polarity of the voltage input connections. The actual circuit analysis of Figure 6-1 is beyond the scope of this textbook. Figure 6-2 shows a panel-mounted E/I converter.

Figure 6-2 Panel-mount E/I converter. *(Courtesy of Moore Industries-International, Inc.)*

There are also current-to-voltage (I/E) transducers in common use in the process control world. A controller, for example, might possibly output current as a signal, while the actuator to which it was transmitting the signal needed a voltage input. Figure 6-3 shows an op-amp circuit that can be used as a current-to-voltage transducer. The transfer function is given by the following equation:

$$E_{out} = R \cdot I_{in} \qquad (6\text{-}3)$$

Figure 6-3 Schematic of an I/E converter.

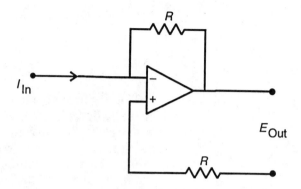

Signal Conditioning. In addition to signal-type conversion, transducers often have circuitry to perform a task known as *signal conditioning.* Signal conditioning is the modification of a signal so that it is transmitted in a more accurate or dependable manner. It includes such tasks as amplification (as mentioned in the thermocouple example earlier), temperature compensation, voltage regulation, noise filtering, and means for adjustment of the span and offset of the current or voltage output range. The tasks just mentioned will now be discussed in turn.

Amplification is often needed because very small signals, especially voltage signals, do not transmit reliably. There are always small resistances located at cable or wire connections or even in the transmitting wires themselves. Very small voltages can be lost because of the voltage drops that will occur across these resistances.

Temperature compensation is often needed because all electrical circuits have some internal heat dissipation and, in addition, are usually exposed to external heat. Heat can change the operating characteristics of the transducer. For example, it might cause an op-amp to heat internally and to increase or decrease its amplification. Temperature compensation helps to ensure that the transducer's output is dependent only on the input.

Voltage regulation guarantees that the output of the transducer will not vary if the power being supplied to operate the transducer varies. For example, nearly all power used to operate instrumentation comes either directly or indirectly from AC line voltage. The AC line voltage can vary slightly due to the changing demand for power during the day by the electric utility's customers. Another possibility is that someone might turn on a large electrical machine somewhere in the plant near the instrumentation equipment. This latter situation could not only lower the amount of power being supplied to the instrumentation area but also introduce considerable electrical noise into the instrumentation system. (*Electrical noise* refers to any electrical voltages that are not supposed to be present, such as voltage spikes.) Voltage regulation would hold the voltage that is operating the transducer (perhaps the voltage being used to power an op-amp) fixed so that the output of the transducer would still be dependent only on the input; and filtering could eliminate the electrical noise created by the nearby machine.

"Span" and "offset" both refer to the output signal range for which the transducer was designed. *Offset* refers to the low end of the range. ("Offset" is very often referred to as "zero.") By doing an offset adjustment, the user is adjusting the transducer so that when the input signal is at the low end of its range, the output signal is at the low end of its range. *Span* refers to the span of the output range, that is, the difference between the high end of the range and the low end of the range. A span adjustment is carried out so that the output signal is at the high end of its range when the input signal is at the high end of its range. (Either adjustment slightly affects the other, so one normally has to go back and forth between making the offset adjustment and the span adjustment until both finally reach their desired values.)

Converters and Transmitters. A further refinement of definitions is due at this point. Since the changing of a signal from voltage form to current form is really not a change of energy form (both are obviously electrical) but, instead, simply a change of variable within the same energy form, the E/I and I/E transducers mentioned in this chapter do not fit the formal definition of transducer as given in Chapter 1. They are more commonly referred to as converters. A *converter* is a device that changes signal representation mode but not the type of energy used as the signal carrier.

It is also somewhat common to denote a transducer or a converter that performs signal conditioning as a *transmitter*. For example, the thermocouple transducer mentioned at the start of the section titled "Transducers" is usually referred to as a thermocouple transmitter. (Figure 6-4 shows a typical thermocouple transmitter.) In practice, however, the three terms—transducer, converter, and transmitter—are often used interchangeably.

136 / SIGNAL TRANSMISSION

Figure 6-4 Panel-mount thermocouple transmitter *(Courtesy of Moore Industries-International, Inc.)*

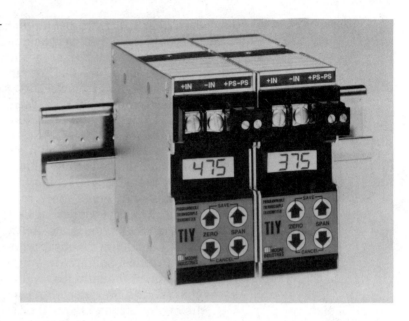

Digital

Binary Transmission. The digital mode of transmission is a two-condition information system that is used because it is relatively easy to detect any erroneous signals that do not fit either of the two conditions. Because it is a two-condition process, digital technology uses the binary number system as illustrated in Figure 6-5. When a system is binary, only numbers that are powers of two are used and only electrical signals that are in one of two possible conditions are used.

Figure 6-5 Conversion table between linear number system and common numbers *(From Jack W. Chaplin,* Instrumentation and Automation for Manufacturing, *© 1992, Delmar Publishers, Inc., Albany, N.Y.)*

Power Value (of 2)	2^8	2^7	2^6	2^5	2^4	2^3	2^2	2^1	2^0
Positive Value (Common Number)	256	128	64	32	16	8	4	2	1

Digital transmission uses electrical signals that are pulses of uniform width. Each pulse represents one binary digit, known as a *bit*. The pulses are allowed to have only two heights or amplitudes, either a "high" amplitude or a "low" amplitude. These two allowable amplitudes are the two-condition requirement referred to in the preceding paragraph. There are several different standardized communication modes used in digital transmission, each with its own high- and low-amplitude criterion. Two of these will be discussed in the next section. For now, it will be assumed that a high pulse is represented by an amplitude of 5.0 volts and a low pulse by an amplitude of 0.5 volt. The low pulse represents a "zero" or "off" state, and a high pulse represents a "one" or "on" state.

Figure 6-6 shows a train of eight pulses or bits being sent from one digital device to another. (Eight bits is defined as a *byte*. Digital information is sent in units of bytes.) The pulses represent, in turn, the first eight powers of the number "2." For example, the first pulse always represents 2^0. If the first pulse is an on state, then 2^0 (which is equal to the common number 1) will be counted. If the first pulse is in an off state, 2^0 will not be counted. The second pulse represents whether the value 2^1 will be added to the value of the first pulse. Again, if the second pulse is an on state, the value 2^1 (which is the common number 2) will be added to the value of the first pulse. If the second pulse is in the off state, nothing will be added to the value of the first pulse. This process would continue until all eight pulses arrive at the receiving end.

Figure 6-6 A byte of data being transmitted in serial mode.

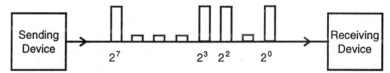

EXAMPLE 6-3

What is the value of the number transmitted from the sending device to the receiving device in Figure 6-6?

Since the pulses representing 2^0, 2^2, 2^3, and 2^7 arrived at the receiving device in the on state, the number transmitted is $1 + 4 + 8 + 128 = 141$.

Of course, using an integer number system such as just outlined is not very useful in sending signals to represent measured variables to, say, four-decimal-place accuracy. However, there are numerous standardized coding forms available in digital systems for sending numbers with many decimal places. The discussion of these systems and of many of the other facets of digital communication would require the use of several textbooks devoted to nothing but digital methods. The discussion of digital methods in this textbook is simply to give the student a rudimentary idea of how digital transmission works.

Standardized Digital Communication Modes. There are several different standards for digital communication. One is the *IEEE-488*. The IEEE-488 standard specifies, among other things, that a high pulse shall be represented by a voltage amplitude of 2.0 v or greater and that a low pulse shall be 0.8 v or less. As mentioned in Chapter 1, the use of allowed amplitude ranges for signals leads to greater accuracy, since a signal can be slightly compromised but still carry the same value (i.e., still be counted as a high bit or a low bit). The IEEE-488 standard is a set of many other rules, specifications, and characteristics for digital communications, but allowable pulse heights are the only ones that are discussed here.

Another commonly used standard is *RS-232*. Among its many specifications, the pulse amplitude range is as follows: Any voltage between −3 v and −25 v represent an "off" control signal, and any voltage between +3 v and +25 v represents an "on" control signal.

A/D Converters and D/A Converters. Sensor/transducers and actuators are usually analog in nature, while controllers are often digital devices. For an analog device to transmit to a digital device, an analog-to-digital converter (ADC or A/D) is needed. Figure 6-7 shows one form of a schematic symbol for an A/D. (If the schematic is reversed right-to-left, it becomes a D/A symbol.) If the eight bits from the A/D are indeed simultaneously transmitted from the A/D, the transmission is called *parallel*. In other words, the communication is conducted by sending series of bytes using parallel communication paths, such as eight parallel wires. If there is only one communication path exiting the A/D, then the bits are being sent sequentially, a communication mode known as *serial* transmission. Electrical pulses can be emitted at such great rates (many thousands per second) that there is often little difference in communication rate between parallel and serial.

In actuality, the schematic diagram in Figure 6-7 is usually not used in the sense that all eight outgoing lines are drawn on a diagram or flowchart, even for parallel output. Too many

Figure 6-7 One type of symbol used for an ADC.

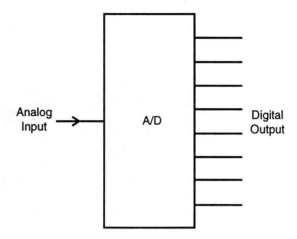

parallel lines make a diagram very cluttered and hide the information that the diagram is trying to show. Usually only one line is used regardless of whether the output is serial or parallel. Figure 6-8 shows a process control loop using an analog sensor, an analog actuator, and a digital controller.

Figure 6-8 A process control loop with an analog sensor, an analog actuator, and a digital controller.

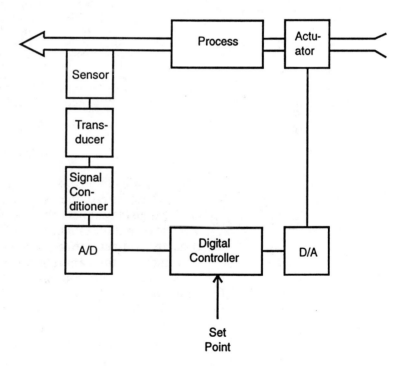

Telemetry

Telemetry (also well known as *telemetering*) is another alternative for transmitting over long distances. It was developed primarily before digital techniques evolved and was at that time the only reliable means for long-distance instrumentation transmission. *Telemetry* is the electrical transmission of information over long distances by any means other than digital.

One method of telemetry is by pulse width. An analog signal of electrical or pneumatic origin is converted to continuously generated pulses, the width of which represent the amplitude of the original analog signal at the instant of conversion. A very simplified outline of a method for converting an analog pneumatic signal to a pulse-width information mode is shown in Figure 6-9.

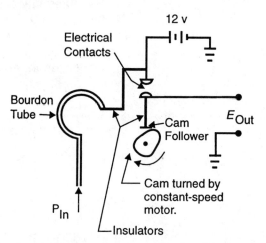

Figure 6-9 Model of a simple telemetry unit for converting pressure to electrical pulse width.

The cam in Figure 6-9 continuously rotates at a constant speed regardless of the value of P_{in}. When P_{in} is at the low end of its range (say, 3 psi), the Bourdon tube will have the upper electrical contact touching the lower contact so that there is constant contact between the two contacts regardless of the movement of the cam follower and the lower contact. Thus, E_{out} will be a continuous signal of 12 v amplitude. This is indicated in Figure 6-10.

As the Bourdon tube senses a greater P_{in} (say, 9 psi), the upper contact is further away from the lower one and they touch less as the cam rotates. Perhaps at this value of P_{in},

Figure 6-10 Output of telemetry unit shown in Figure 6-9.

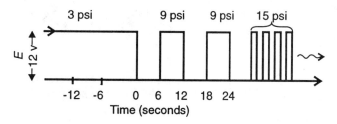

they touch during only one-half of a cam revolution. If the cam rotates at one revolution per six seconds, there will be pulses of 12 v amplitude and a three-second width produced. This is shown in the central portion of Figure 6-10. Now assume that P_{in} rises to the upper end of its range (say, 15 psi) and that the contacts now stay closed only one-sixth of a cam revolution. This will result in pulse widths of one second. At the other end of the transmission line, a reverse converter is needed to convert pulse width to whatever variable the controller is set up to operate with—pneumatic, current, voltage, and so on.

Notice that this form of transmission does not depend on voltage amplitude. The voltage can be compromised by partial shorts, by electrical noise, and the like; but it is only the width of the pulse that conveys information.

Another method of telemetry is to convert analog amplitude data to frequency data. The result of a voltage-to-frequency conversion is illustrated by Figure 6-11. Notice that the amplitude of the electrical signal is constant. It is only the frequency that is varying. The receiving end would probably be using a frequency-to-voltage converter.

Figure 6-11 Output of a voltage-to-frequency telemetry unit.

PNEUMATIC

Standardized Ranges. Pneumatic transmission is most commonly used in the heating, ventilation, and air conditioning (HVAC) industry, is always analog, is useful for only short distances, and is standardized to two ranges: 3 to 15 psi or 6 to 30 psi. Notice that neither range begins at zero. The reason for not using the value "0" for the low end of the ranges is twofold: (1) low pressures do not transmit very well without being either lost or severely degraded, and (2) by using a non-zero end value, the operator can determine when a system is down. If "0" represented the low end of the range, the operator would not know if the system was down or whether simply a signal representing the very low end of the range was being transmitted.

Pneumatic Transducers. Transducers to convert pressure to current (P/I), pressure to voltage (P/E), and the opposite of each (I/P and E/P) are very common. An example of how a simple P/E transducer might be constructed is shown in Figure 6-12.

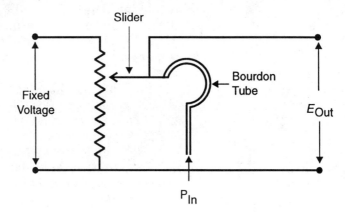

Figure 6-12 Simple model of a P/E transducer.

FIBER OPTICS

Another mode of transmission that is being used more frequently throughout the communication world (not just the industrial communication world) is fiber optics. *Fiber optics* is the transmission of information through the use of light signals traveling through specialized optical paths. It is actually only the transmission paths that are different insofar as process control is concerned. At the start and end of each path, a fiber-optic/electrical interface is used and everything else stays the same in regard to the process control system elements. Figure 6-13 shows a fiber-optic/digital transducer.

Figure 6-13 Fiber-optic/digital electronics transducer *(Courtesy of EOTec Corporation; from Jack W. Chaplin,* Instrumentation and Automation for Manufacturing, *© 1992, Delmar Publishers, Inc., Albany, N.Y.)*

REVIEW MATERIALS

Important Terms

transmission
signal conditioning
offset
converter
bit
on state
IEEE-488
parallel transmission
telemetry

transducer function
electrical noise
span
transmitter
byte
off state
RS-232
serial transmission
fiber optics

Questions

1. What forms of transmission are not normally used over long distances, where "long distance" is defined as any distance over approximately 1,500 feet?
2. Why are standardized ranges for signal values useful?
3. List all of the standardized ranges that you can think of, including the type of variable for which they exist.
4. Are current and voltage different types of energy, or are they different forms of the same type of energy?
5. Can digital equipment and analog electrical equipment both be used in the same process control system? Discuss.

6. List some of the functions or results of signal conditioning.
7. How many adjustments does it take to calibrate the span and the offset of a transducer?
8. How does a converter differ from a transducer? From a transmitter?
9. If a digital transmitter sent thirty-two bits serially to a receiver, how many bytes of information will have been transmitted?
10. Would a +2 v pulse represent an on state or an off state under RS-232 standards?
11. Which form of digital transmission would be faster (all other factors equal)—serial or parallel?
12. How does telemetering differ from telemetry?
13. Suppose the rate of revolution of the cam in Figure 6-9 was doubled. What would happen to the pulse widths shown in Figure 6-10? What would happen to the pulse amplitudes?
14. What are the advantages of not using a zero value as the low end of a standardized transmission range?
15. If the transmission line from the A/D to the controller in Figure 6-8 was a fiber-optic line, what else would need to be added to the block diagram?

Problems

1. What would be the final resulting voltage of a 10 v, 10 mA signal sent over a 300 ft, 22 gage wire, if the resistance of the wire per foot was 0.01 Ω? Over a 1,500 ft wire?
2. What would be the result for Problem 1 if a 1 v, 1 A signal was sent, instead of the 10 mA signal?
3. If a thermocouple outputs a 560 μv signal at the maximum end of its range, what amplification would be needed for the transmitter to output a 10 v signal?
4. It has been decided that the output signal from the transmitter in Problem 3 would be better if it was in current form. What ratio of resistors would be needed to convert the 10 v input (to the final stage of the transmitter) to a 20 mA output?
5. For Problem 4, use some values for R_1, R_2, and R_3 in the range of 100 Ω and the range of 1,000 Ω and arrive at some suitable values for the three resistors. (Note that there can be many different answers to this problem.)
6. After working Problem 5, decide on some workable values for R_5 and R_4.
7. Suppose it was desired to use an I/E transducer to convert 20 mA to 5 v? What values would be needed for the two resistors in Figure 6-3?
8. Draw a signal representation diagram similar to Figure 6-6 to represent the transmission of the number 133.
9. Using the RS-232 format, draw a signal representation diagram to represent the transmission of the number 87.

10. If the IEEE-488 format were being used and the following eight pulses were received at the controller, what number will have been transmitted? (The pulses were received in order; that is, the 5.0 v pulse was received first.) (a) 5.0 v, 5.0 v, 4.9 v, 0 v, 0 v, 0.1 v, 4.9 v, 0 v; (b) 5.0 v, 0 v, 0 v, 0.1 v, 4.9 v, 1.9 v, 5.0 v, 4.9 v

11. Redraw Figure 6-10 if the cam revolves at the rate of 30 revolutions per minute. Use the same time scale. If the value of E was cut in half, how would this affect the information being transmitted?

12. Reproduce Figure 6-11 and, assuming that the voltage-to-frequency converter has $f_{out} = kE_{in}$ as its transfer function and where k is a positive constant, identify the regions representing the higher-voltage amplitudes.

13. What would be the value of k in Problem 12 (including units) if, for $E_{in} = 10$ v, it was desired that $f_{out} = 600$ Hz?

14. Modify Figure 6-12 so that it would perform as a simple P/I transducer.

Temperature and Heat

CHAPTER GOALS

After completing study of this chapter, you should be able to do the following:

- Understand the difference between temperature and heat.
- Work with the different temperature scales and convert readings from one scale to another.
- Understand the phase behavior of materials.
- Calculate the heat required to increase the temperature of a given amount of material by a specified number of degrees.
- Understand the heat-transfer mechanisms of conduction, convection, and radiation.
- Calculate the amount of heat conducted under a specified set of conditions.
- Understand the concept of thermal expansion and make calculations relating to linear and volumetric expansion.
- Understand the operation of the different types of thermometers.
- Describe what thermistors, thermocouples, and pyrometers are and how they work.
- Understand the concept of a time constant and make appropriate claculations,
- Apply what you learn to selecting, installing, calibrating, and protecting temperature-measuring devices.

148 / TEMPERATURE AND HEAT

> Various temperatures affect our daily lives, including body temperatures, which can be indicators of our health, and air temperatures, which have to do with our comfort. With regard to industrial activity, temperature is probably the most common and important of the variables. Heating and cooling with the associated temperature changes are involved with both physical and chemical changes in materials.

BASIC CONSIDERATIONS

Basic Terms

Temperature and Heat. The hotness or coldness of an object is indicated by its *temperature*. *Heat* is a form of energy; and when heat is applied to an object, the molecules making up the object become more active. This increase in molecular activity results in an increase in temperature. Temperature is measured in *degrees* on one of several different *scales*. Two of the most common measurements for heat are the *BTU (British Thermal Unit)* and the *calorie*.

Fahrenheit Scale. One of the first scales to be widely used for measuring temperature was proposed in the early 1700s by a Dutch instrument maker named Fahrenheit. He chose the value of 0 degrees on his scale as the temperature of a mixture of water, salt, and ice. Later, the two most common references on his scale were set at 32 degrees, the freezing point of water at atmospheric pressure, and 212 degrees, the boiling point of water at atmospheric pressure. Thus, there are 180 degrees between the two reference points.

Celsius Scale. Another widely used temperature scale was devised in the mid-1700s by a Swedish astronomer named Celsius. On this scale, the freezing and boiling points for water at atmospheric pressure are 0 degrees and 100 degrees, respectively. The difference between the two references is thus 100 degrees. There was a period of time when this temperature scale was called the centigrade scale. That term is now seldom used.

Absolute Scales. Theoretically, there is a condition where all molecular motion ceases. The temperature associated with this condition is referred to as *absolute zero*. Laboratory experiments have been conducted in which this temperature has been approached. The behavior of materials usually changes markedly in the vicinity of this temperature.

Two temperature scales have been devised which take absolute zero into account. The absolute scale related to the Fahrenheit scale is named after W. J. M. Rankine, who performed scientific and engineering work in the mid-1800s. The size of the Fahrenheit and Rankine degrees are the same; that is, an increase of 1° F is the same as an increase of 1° R. The Rankine temperature corresponding to 0° Fahrenheit is approximately equal to 460° R. See Figure 7-1a for a Rankine–Fahrenheit comparison.

The absolute scale related to the Celsius scale was named after W. Thompson, also known as Lord Kelvin, who was involved with scientific work in the mid- to late 1800s. The size of the Celsius and Kelvin degrees are the same; that is, an increase of one Celsius degree is the same as an increase of one Kelvin degree. The Kelvin temperature corresponding to 0° Celsius is approximately 273° K. See Figure 7-1b for a Kelvin–Celsius comparison.

British Thermal Units. The *British Thermal Unit (BTU)* is one of the most frequently used units for measuring quantities of heat. By definition, a BTU is the amount of heat required to raise the temperature of 1 pound of water 1° F at 68° F and atmospheric pressure. In terms of an equivalent amount of energy, 778 foot-pounds are equal to 1 BTU.

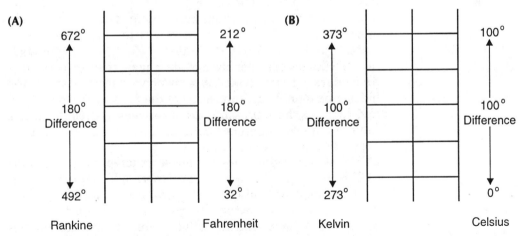

Figure 7-1 A comparison of temperature scales: **(A)** Rankine–Fahrenheit; **(B)** Kelvin–Celsius.

Calories. The *calorie* is another widely used heat unit. One calorie is defined as the amount of heat required to raise 1 gram of water 1° Celsius. In terms of SI units, 4.19 calories is approximately the equivalent of 1 joule. For purposes of comparing BTUs and calories, 1 BTU is equal to 252 calories.

Table 7-1 gives unit conversions related to heat.

Table 7-1 Unit Conversions Related to Heat Energy

1 BTU	=	252 calories
1 calorie	=	0.0039 BTU
1 BTU	=	1,055 joule
1 joule	=	0.000948 BTU
1 BTU	=	778 foot-pounds
1 foot-pound	=	0.001285 BTU
1 calorie	=	4.19 joules
1 joule	=	0.239 calorie
1 foot-pound	=	0.324 calorie
1 joule	=	0.738 foot-pound
1 foot-pound	=	1.355 joules

Change of Phase. In general, matter can exist in one of three phases: *solid, liquid,* or *gas*. Probably the most familiar example is water, which occurs in daily life in the solid phase (ice), liquid phase, and gaseous phase (steam). Whenever a substance undergoes a change of phase, heat is either absorbed or liberated and there is usually a change in the volume occupied by the substance. For example, again referring to water, heat must be added to ice to melt it into a liquid, which, in turn, must be heated to obtain steam. The reverse operation, removal of heat, is necessary to condense steam to water and then to freeze the water to ice. In heating or cooling substances through changes in phase, there are two distinct operations involved. If heat is applied to ice at 32° F, the ice begins to melt. The water from the melting ice is at 32° F and remains there until the ice is completely melted. At this time, continued application of heat will then increase the temperature of the water. When the water temperature reaches 212° F, steam begins to form. The water temperature remains 212° F; and after all the liquid is evaporated, continued application of heat will raise the temperature of the steam above 212° F. The reverse situation, going from steam to ice, requires the removal of heat both to reduce the steam and liquid temperature and to pass through the two changes in phase.

The relationship between the addition of heat to a substance and the changes in phase that take place is shown in Figure 7-2.

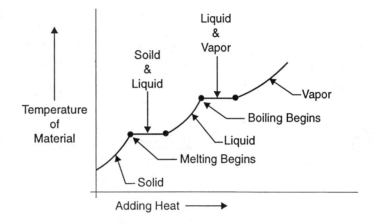

Figure 7-2 Matter undergoes changes of phase as heat is added or removed.

It should be noted that there are situations in which the change in phase bypasses the liquid stage. When a solid changes phase directly to a gas, the process is called *sublimation*.

Specific Heat. *Specific heat* is defined as the amount of heat required to raise a definite amount of a substance by one degree. In engineering units, it is usually given as BTUs per pound per degree Fahrenheit. That means that if a material had a specific heat of 0.5 BTU/lb/°F, it would take 0.5 BTU to raise 1 pound of the material 1° F. In terms of calories, specific heat is given in calories per gram per degree Celsius. Table 7-2 gives specific heat values for some common materials.

Thermal Conductivity, Convection, and Radiation. Heat flows from regions of higher temperature to regions of lower temperature. *Heat transfer* is the study of heat movement. There are three basic heat-transfer mechanisms: conduction, convection, and radiation. Although these mechanisms can be considered separately, many practical applications involve two of them and sometimes all three.

Conduction involves the movement of heat from one molecule of an object to another. Thus, if one end of a metal rod is at a higher temperature than the other end, heat will be transferred by conduction to the cooler end. The *thermal conductivity* of a material is a measure of its ability to conduct heat and is usually designated by k. The units are BTU per

Table 7-2 Specific Heats for Some Common Materials (BTUs per pound per Fahrenheit degree or calories per gram per Celsius degree)

Alcohol	0.58 to 0.60
Aluminum	0.214
Brass	0.089
Cast iron	0.119
Copper	0.092
Glass	0.12 to 0.16
Gold	0.0316
Lead	0.031
Mercury	0.033
Platinum	0.032
Quartz	0.188
Silver	0.056
Steel	0.107
Tin	0.054
Water	1.000

Table 7-3 Thermal Conductivities for Some Common Materials (BTUs per hour per foot per Fahrenheit degree)

Air	0.016 (room temp.)
Aluminum	119
Brass	52
Brick	0.40
Concrete	0.80
Copper	220
Mercury	4.8
Silver	242
Steel	26
Water	0.36 (room temp.)

hour per foot per degree Fahrenheit. Some typical thermal conductivity values are given in Table 7-3. The thermal conductivities of most materials vary at least slightly with temperature. In a more extreme case, the k value for air approximately doubles between 300 and 1,300° F.

Convection is the term applied to the transfer of heat by the motion of warm or hot material. Hot-air furnaces and hot-water heating systems are examples of heat transfer by convection. If the material moves strictly because of differences in density, the process is called *free* or *natural* convection. Warm air rising in an otherwise quiet room is being convected. If pumps, blowers, or fans are used to move the material, the process is called *forced* convection.

Radiation refers to the emission of energy from a body. The energy is in the form of electromagnetic waves which travel at the speed of light and are transmitted through any medium (solid, liquid, or gas) or vacuum. Probably the best example of radiant heat transfer is the heat we receive from the sun.

As a further example of the three heat-transfer mechanisms, consider a warm object sitting on the floor in a room. A person placing their hand on the object would feel heat conducted to their hand from the surface of the object. If the person then held their hand slightly above the object, they could feel the warm air rising (convecting) because it is "lighter" than the surrounding cooler air. Finally, the person could hold their hand slightly away and to the side of the object. Although there would be a slight amount of conduction from the object surface to the air and then through the air to the hand, radiation would be the primary reason for the warmth felt.

Properties and calculations involved with convection and radiation tend to be more involved than with conduction.

Thermal Expansion. Almost all materials expand when heated. The amount of expansion is related to the *coefficient of thermal expansion*. These coefficients can be for *linear* or *volumetric* changes. Coefficients of expansion are usually given as the amount of expansion per degree temperature change, with the linear coefficient designated as α and the volumetric coefficient as β. Table 7-4 lists values for some common materials.

Formulas Related to Temperature and Heat

It is frequently necessary to convert readings from one temperature scale to another. The four following sections provide the formulas for doing so.

Table 7-4 Thermal Coefficients of Expansion for Some Common Materials (linear and volumetric expansion per Fahrenheit degree)

	Linear	Volumetric
Alcohol	—	0.00061 to 0.00066
Aluminum	0.0000128	—
Brass	0.00001	—
Cast iron	0.0000056	0.000020
Copper	0.0000094	0.000039
Glass	0.000005	0.000014
Gold	0.0000078	—
Lead	0.000016	—
Mercury	—	0.0001
Platinum	0.000005	0.000015
Quartz	0.00000022	—
Silver	0.000011	0.000032
Steel	0.0000061	—
Tin	0.000015	0.000038

Fahrenheit and Celsius. The formula for converting a Fahrenheit reading to a Celsius reading is

$$°C = \frac{5}{9}(°F - 32) \tag{7-1}$$

Using this formula, a reading of 212° F would become

$$°C = \frac{5}{9}(212 - 32) = 100° \tag{7-2}$$

To go from a Celsius reading to a Fahrenheit reading:

$$°F = \frac{9\,(°C)}{5} + 32 \tag{7-3}$$

Using this formula, a reading of 100° C would become

$$°F = \frac{9\,(100)}{5} + 32 = 212° \tag{7-4}$$

Fahrenheit and Rankine. For converting a Fahrenheit temperature to an absolute temperature on the Rankine scale:

$$°R = °F + 460 \tag{7-5}$$

Thus, a temperature of 100° F would be the equivalent of 560° R.

Celsius and Kelvin. For converting a Celsius temperature to an absolute temperature on the Kelvin scale:

$$°K = °C + 273 \tag{7-6}$$

A Celsius temperature of 40° C would be 313° K.

Rankine and Kelvin. Occasionally, it is necessary to convert readings from one absolute temperature scale to the other. This can be achieved with the following formulas:

$$°R = 1.802° K \tag{7-7}$$

$$°K = 0.555° R \tag{7-8}$$

EXAMPLE 7-1

The freezing temperature for water is 273° K. Use the previous equations to convert this reading to Rankine and then Fahrenheit readings.

$$°R = 1.802 (273) = 492°$$

$$492 = °F + 460$$

$$°F = 32$$

Amount of Heat. The amount of heat required to raise or lower the temperature of a given amount of material can be calculated using the following formula:

$$Q = WC (T_2 - T_1) \tag{7-9}$$

where

Q = amount of heat required

W = weight of material

C = specific heat of the material

T_2 = final temperature

T_1 = starting temperature

EXAMPLE 7-2

A substance has a specific heat of 0.3 BTU/lb °F. It is desired to heat 2 pounds of the substance from 50° F to 80° F. How much heat is required?

$$Q = (2 \text{ lbs}) \, 0.3 \text{ BTU/lb °F} \, (80 - 50) \text{ °F}$$

$$Q = 1.8 \text{ BTUs}$$

Notice that if the final temperature were less than the starting temperature, the answer would be negative, indicating that heat would have to be removed from the material. Use of Equation 7-9 requires diligence to ensure that the proper units are being used.

Heat Conduction. The amount of heat conducted in a given application can be calculated using an equation of the following form:

$$Q = -\frac{kA(T_2 - T_1)}{L} \qquad (7\text{-}10)$$

where

Q = rate of heat transfer

k = thermal conductivity of the material through which heat is moving

A = cross-sectional area through which heat is moving

T_2 = temperature of material at point toward which heat is moving

T_1 = temperature of material at point from which heat is moving

L = length of path between points of temperature measurement

Again, care must be taken to keep consistent units.

EXAMPLE 7-3

How much heat is transferred per hour through 6 square feet of a brick wall if one side is at 100° F and the other side is at 70° F? The wall is 4 inches thick. The thermal conductivity of the brick is 0.40 BTU/hr ft °F.

$$Q = -\frac{0.40\,(6)\,(70-100)\,°F}{0.333}$$

$$= 216 \text{ BTUs/hour}$$

Note that because T_2 is less than T_1, the temperature difference is negative and the result is a positive heat flow.

Heat Convection and Radiation. Calculations for convection and radiation are usually not as straightforward as for conduction. A typical heat convection formula is of the following form:

$$Q = hA\,(T_2 - T_1) \qquad (7\text{-}11)$$

where

Q = convection heat transfer rate

h = coefficient of heat transfer

A = heat transfer area

$T_2 - T_1$ = temperature difference between the surface and the bulk of the fluid away from the surface

Although Equation 7-11 is simple in form, making the proper choice for a value of "h" can be difficult. For both natural and forced convection, equations and charts are available which take into account temperatures, distances, densities, viscosities, specific heats, and so on. Information is available for horizontal and vertical plates and pipes. Previous experience can be very valuable in selecting "h" values.

EXAMPLE 7-4

A 10 foot by 10 foot surface has a coefficient of heat transfer of 0.4 BTU/hr ft² °F. How much heat is transferred if the temperature difference is 50° F?

$$Q = \frac{0.4 \text{ BTU}}{\text{hr ft}^2 \text{ °F}} (100 \text{ ft}^2)(50° \text{ F})$$

$$= 2{,}000 \text{ BTUs/hour}$$

A basic type of radiation heat transfer is of the following form:

$$Q = CA\,(T_1^4 - T_2^4) \tag{7-12}$$

where

Q = heat transferred

C = a constant depending on units used

A = area of the radiating surface

T_1 and T_2 = absolute temperatures of the surfaces involved

This formula is for the radiation between two bodies. To make specific calculations, more information is required about the geometry of the surfaces and the physical properties of the surface materials.

EXAMPLE 7-5

A heating unit has a surface area of 40 ft². If its surface temperature is 100° F, how much heat is radiated if the walls of the room are at 70° F? Consider the constant involved to be 0.173 × 10⁻⁸ BTU/hr ft² °F⁴.

$$Q = 0.173 \times 10^{-8} \text{ BTU/hr ft}^2 \text{ °F}^4 \, (40 \text{ ft}^2)$$
$$[(100 + 460)^4 - (70 + 460)^4] \text{ °F}^4$$

$$= 1{,}340 \text{ BTUs/hour}$$

Thermal Expansion. Linear expansion is calculated using the following formula:

$$L_2 = L_1 [1 + *(T_2 - T_1)] \qquad (7\text{-}13)$$

where

L_2 = final length

L_1 = initial length

α = coefficient of linear thermal expansion

T_2 = final temperature

T_1 = initial temperature

Use of consistent units is necessary for meaningful results. Note that the "$\alpha(T_2 - T_1)$" part of the formula leads to the increase (or decrease in the case of a decreasing temperature) in length.

The change in volume of a material is calculated in a similar manner:

$$V_2 = V_1 [1 + \beta(T_2 - T_1)] \qquad (7\text{-}14)$$

where

V_2 = final volume

V_1 = initial volume

β = coefficient of volumetric thermal expansion

T_2 = final temperature

T_1 = initial temperature

As can be seen in Table 7-4, the value of β is usually about three times that for α.

EXAMPLE 7-6

A platinum rod is 20 inches long and has a diameter of 0.5 inch. Calculate its new length and volume for a temperature increase of 200° F.

New length = 20 [1 + 0.000005 (200)]

= 20.02 inches

New volume = $\dfrac{\pi (0.5)^2}{4}$ (20) [1 + 0.000015 (200)]

= 3.939 in^3 inches (original volume was 3.927 in^3)

MEASURING DEVICES

The measurement of temperature is usually based on one of the following physical properties:

1. Expansion of a material resulting in a change in length, volume, or pressure.
2. Changes in electrical resistance.
3. Changes in contact voltage between different metals.
4. Changes in radiated energy.

Following are some of the particular applications.

Thermometers

Thermometer is the general name used to describe a group of instruments used to measure temperature (see Figure 7-3). Several of the types available are discussed here.

Mercury. Mercury has long been used for measuring temperature. Probably the most common application is the *mercury-in-glass* thermometer. Mercury is placed in a small-bore glass tube that is then sealed. When the base or *bulb* of the thermometer is heated, there is thermal expansion of both the glass and mercury; but since the coefficient of expansion is larger for mercury, it rises in the tube. The amount of rise

160 / TEMPERATURE AND HEAT

Figure 7-3 Thermometers
(Courtesy of Cooper Instrument Corporation)

can be correlated with the increase in temperature. Mercury-in-glass thermometers can usually be used in the general range of –30° F to 800° F.

Liquid. *Liquid-in-glass* thermometers operate in a manner similar to mercury-in-glass (mercury is limited because of its freezing point of approximately –38° F). For use in a liquid-in-glass thermometer, the following are preferable properties:

1. The liquid should have a large coefficient of expansion to make readings more accurate.
2. The liquid expansion should be linear with temperature.
3. The liquid should be easily visible even if present in a small cross-sectional area.
4. The liquid should not stick to the glass walls.

Liquid thermometers can be made in many sizes. The range of temperatures they cover is usually small, but the accuracy is usually good. A general application range would be from –300° F to 600° F.

Bimetallic. Different metals expand at different rates when heated. This phenomenon is applied in *bimetallic* thermometers. Two metals are joined along their length, Figure 7-4. When the assembly is heated, the free end moves in the direction of the more slowly expanding metal. This deflection can be correlated with temperature change. The bimetallic strip is usually formed into a spiral or helix to fit in a gage. A pointer then indicates the temperature. A general application range for bimetallic thermometers is −300° F to 800° F. One of the most common application of bimetallic thermometers is in home-heating thermostats.

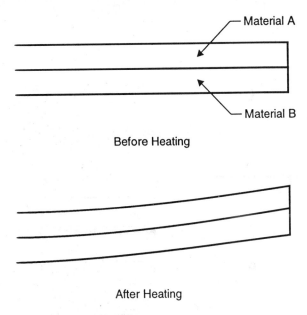

Figure 7-4 Principle of the bimetallic thermometer—material B has a higher expansion coefficient than material A.

Pressure-Spring. Glass and bimetallic thermometers are suitable for measuring temperatures for local application. Often, it is desired to read and record the temperature away from the measurement site. Pressure-spring thermometers are suitable for this purpose. The basic component is a tube which is filled with a pressure-sensing substance. Changes in the temperature of the substance cause it to expand or contract; and since the tube is closed, the pressure rises. The tube is formed in the shape of a Bourdon tube, a spiral, or a helix (see Chapter 5). One end of the tube is joined to a pressure spring, which is converted through a pointer into a gage reading; the other end is joined to a sensing bulb which is the part of the thermometer that contacts the substance whose temperature is to be measured.

There are four basic classes of pressure-spring thermometers:

Class 1   Liquid-filled (excluding mercury)
Class 2   Vapor pressure
Class 3   Gas-filled
Class 4   Mercury-filled

The measuring range of a liquid-filled device is determined by the volume of liquid in the bulb; the larger the volume, the wider the range. There is a possibility of errors being introduced because of temperature variation along the tube or at the spring. The effect of these errors can be minimized using special compensating devices (see Figure 7-5).

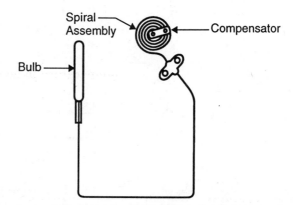

Figure 7-5 Bulb and compensating assembly for temperature measurement.

Vapor pressure devices (Figure 7-6) depend on the vapor pressure of a liquid that only partially fills the system. The liquid can expand when heated; but as it does, its vapor pressure increases. Because the vapor pressure does not increase linearly (increases in pressure are not proportional to increases in temperature), the scale divisions on the gage will not be uniformly spaced unless special provisions are made.

Another potential problem with a vapor pressure system can occur because of the movement of gage fluid from the liquid phase to the gaseous phase or vice versa. If this takes place, temperature readings can be affected. Most manufacturers use a method of constructing the device to bypass this

Figure 7-6 Vapor thermometer *(Courtesy of Cooper Instrument Corporation)*

problem. Two different fluids are used—one vaporizing and one nonvaporizing—in this procedure.

Selecting fluids for a vapor pressure system involves the following:

1. The fluid must not react with the bulb or pressure element.
2. The boiling point of the fluid must be below the lowest temperature to be measured so that the vapor pressure can develop sufficiently.
3. The highest temperature to be measured must be less than the critical pressure of fluid. Common fluids used in vapor pressure systems include ethyl alcohol, ether, toluene, and methyl chloride.

Gas-filled systems are based on the fact that, because they have a constant volume of gas, the temperature changes take place in direct relationship to the pressure changes. This effect can be shown as a form of Charles's law:

$$\frac{\text{initial pressure}}{\text{final pressure}} = \frac{\text{initial temperature}}{\text{final temperature}}$$

where both the pressure and temperature must be on the absolute scales. Gas-filled systems are usually filled under high pressure (150 to 500 psi at room temperature) to obtain a greater increase in pressure for a given temperature rise. Nitrogen is commonly used because of its relative chemical inactivity and favorable expansion rate. Gas-filled systems are much like liquid-filled systems except for bulb size. The gas-filled bulb must be larger so that its volume will be larger than the volume of the rest of the system.

Class 4, mercury-filled systems, are much like liquid-filled systems (see previous discussion). Mercury-filled systems are suitable for measuring temperatures up to approximately 1,000° F.

Resistance. The electrical resistance of almost all pure metals increases with temperature. The amount of resistance change is uniform with temperature change. This makes them appropriate candidates for temperature measurement. Construction usually consists of a bulb formed by wrapping a fine wire around an insulator and enclosing it in metal. Considerations in wire selection include the following:

1. High resistance change per degree temperature change
2. Metal purity and uniformity
3. Stability
4. Contamination resistance

Metals used include platinum, nickel, and copper. An approximate range for resistance thermometers is –300° to 1,200° F. Typical application involves the use of a Wheatstone bridge arrangement. In some applications, care must be taken so that heat generated by electrical current flow does not affect the temperature readings.

Thermistors. Some metal oxides experience a *decrease* in electrical resistance as their temperature increases. Although the rate of change is not as uniform as the resistance of pure metals, these materials have found application in temperature measurement and are called *thermistors* (see Figure 7-7). Thermistors can be made in a variety of shapes to fit specific applications. They usually have a fast response and high sensitivity. Common materials include oxides of cobalt, nickel, and manganese. Thermistors are used more commonly in lower temperature ranges and, like resistance thermometers, typically in a Wheatstone bridge arrangement.

Thermocouples. When the ends of two dissimilar metals are joined as in Figure 7-8, the result is a simple *thermocouple*. If the two ends (junctions) of the thermocouple are at different

Figure 7-7 Several types of thermistors *(Courtesy of Fenwal Electronics, Inc.)*

temperatures, an electrical current will flow through the circuit formed by the two wires. The magnitude of the current can be related to the temperature difference between the two ends. Although a thermocouple is a temperature-difference measuring device, it can be used as a temperature-sensing device if one of its ends is kept at a constant temperature. A common application uses an ice–water combination (32° F) as a temperature source for one end of the thermocouple. The current flowing then indicates the amount that the second end is above or below 32° F.

The thermoelectric phenomena involved in the operation of a thermocouple are named after three of the people involved in their study. The Seebeck effect says that the voltage produced varies with the temperature difference between the two junctions. The Peltier effect involves the fact that if a current flows across the junction of two dissimilar metals which have the same temperature, heat is released or absorbed. The Thomson effect results when an electric current passes

Figure 7-8 Basic set-up for a thermocouple; wire A and wire B are made of different metals.

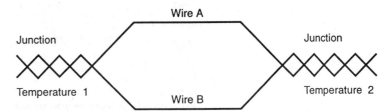

through a conductor along which the temperature varies. Heat is again released or absorbed, depending on the direction of the current and direction of the heat variation. The voltage which produces the current (Seebeck effect) is the sum of the electromotive forces generated at the junctions (Peltier effect) and along the dissimilar metals (Thomson effect).

Measuring the electrical potential across the junctions of a thermocouple turns out to be more convenient than measuring the current. In the development of a thermocouple circuit, several "thermocouple laws" may be involved. The first law says that the thermoelectric effect depends only on the temperatures of the junctions and is not affected by temperatures along the wires. The practical effect of this law is that the temperature of connecting leads does not enter into thermocouple readings. This law is sometimes called the law of intermediate temperatures.

The effect of the second law is that metals other than those making up the thermocouple can be used in the circuit without affecting the potential as long as the junctions of the additional metals are at the same temperature. Running of wires through terminal strips or connectors is thus appropriate. The third law comes from the first and provides that a third metal can be introduced at either of the two original junctions as long as the temperature at both ends of the third wire is the same. This allows for connecting to measuring devices.

Sometimes called the law of intermediate metals, the fourth law allows calculation of the expected results for a combination of two particular metals if each of their outputs with a third metal is known. The fifth law is related to a basic equation involving the potential and the absolute temperature. Its application is involved with interpolating between values given in a thermocouple table.

Table 7-5 lists sample operating temperatures for thermocouples. Note that they can be used in temperatures exceeding 3,000° F. Figure 7-9 shows typical thermocouples.

Table 7-5 Typical Operating Ranges for Thermocouples

Type	Approximate Range, °F
Copper-constantan	−250 to 700
Chromel-constantan	−320 to 1,800
Iron-constantan	32 to 1,600
Chromel-alumel	32 to 2,500
Platinum-platinum-rhodium	32 to 3,000

Figure 7-9 Five types of industrial thermocouples *(Courtesy of ARI Industries, Inc.)*

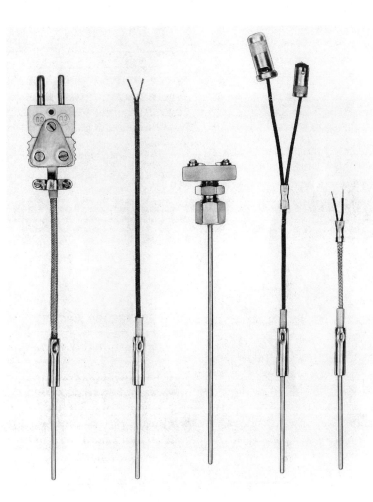

Pyrometers

A *pyrometer* is a device that measures the temperature of an object by "viewing" it through a fixed lens (Figure 7-10). Radiation emitted by the object passes through the lens and impacts on the temperature-sensing device. Usually, the sensing device is a circular ring of thermocouples connected in series (known as a *thermopile* [see Figure 7-11]). Since the object whose temperature is to be measured emits radiation in all directions, it is sufficient that it be in the field of view of the sensor. It can be located at any distance, provided it fills the field of view of the sensor.

Figure 7-10 Principle of pyrometer operation—the lens focuses heat rays on the thermopile.

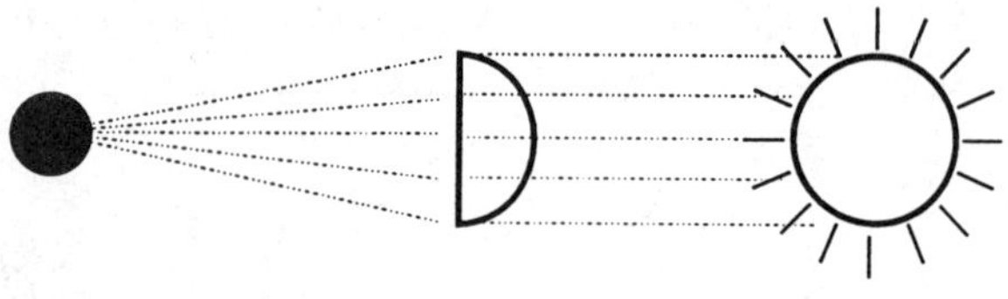

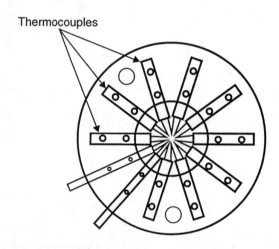

Thermopile   Lens   Heated Object

Figure 7-11 (*right*) Thermopile showing thermocouples.

Figure 7-12 Industrial hand-held pyrometer *(Courtesy of Capintec Instruments, Inc., Ramsey, N.J.)*

In another type of pyrometer, the color of a glowing wire is compared with the color of the object of interest. The temperature is then obtained using the measured electric current. A hand-held pyrometer is shown in Figure 7-12.

APPLICATION CONSIDERATIONS

Selection

Selection of a device for measuring temperature is based on finding a device which measures properly for the given temperature conditions and is compatible with the surrounding environment. In the following, several factors are discussed which impact on the selection process.

Bulb Size. For thermometers, the size of the bulb containing the fluid has two direct effects. A larger bulb contains more fluid and thus can cover a larger span of temperatures. It also, however, takes longer to change from one reading to another, since there is more fluid to heat.

Time Constant. The *thermal time* constant for a device is defined as the time required for it to change 63.2% of the difference between two different temperatures. If a thermometer were in a 0° F water-ice combination long enough for it to register 0° F and then suddenly moved to a location where the temperature was 100° F, the time constant would be the time required for it to reach 63.2° F. The rate of change is exponential, which means that in the next amount of time equal to the time constant, the thermometer would climb 63.2% of the remainder of the way, or 0.632 (100 − 63.2) = 23.3° F. At the end of the time equal to two time constants, the thermometer would read 63.2 + 23.3 = 86.5° F. Continuing this development the thermometer will read 99.9% of the true value after a period equal to seven time constants (see Figure 7-13). It should be noted that there are some temperature-sensing devices that do not follow this exponential law.

Figure 7-13 Response time for two temperature measuring devices.

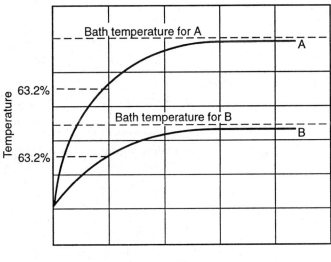

EXAMPLE 7-7

Continuing with the description in the previous paragraph, calculate the temperature reading after the amount of time equal to three time constants.

At the end of two time constants, the reading was 86.5° F. The increase would be 0.632 (100 − 86.5) = 8.5° F. Thus, at the end of the third period the temperature would be 86.5 + 8.5 = 95°.

The *range* of a temperature-sensing device indicates the lowest and highest readings it can make. The *span* gives the difference between these two readings. Thus, a device capable of measuring between –50° F and 350° F would have a span of 400°, with the range given by the high and low values.

EXAMPLE 7-8

The accuracy of a particular temperature-sensing device is given as ±1% of its span. If its range is 50° F to 500° F, what would be the probable upper and lower bounds for a reading of 350° F?

Span = 500° F – 50° F = 450° F

1% of 450° = 4.5°

Bounds would be 350° F ± 4.5°, or 345.5° F and 354.5° F

Some temperature-sensing devices work well in particular environments. Bimetal devices can withstand vibration. Thermistors have been applied in overload protectors, warning devices, temperature control, and the like.

Installation

Bulb Location. The sensing portion of the temperature-measuring device should be located so as to react only to the temperature of the material or process of interest and not to reflected heat. Care must be taken so that the sensing portion does not touch such objects as cold metal or walls. The sensing portion should always be placed as close to the first point at which the measurement can be made so as to enable early readings for observation and control. When measuring the temperature of fluids being mixed, the sensing portion must be far enough downstream to ensure that the mixture temperature has stabilized.

Compensation. Compensation is necessary in some long systems. In these systems, using liquid- or mercury-filled thermometers, various methods can be applied, including a second pressure spring, special tubing, or a bimetallic strip. Although compensation requires extra equipment and time, it should be carefully planned and implemented when accurate readings are required.

Calibration

Calibration of glass thermometers can be as simple as "pointing out" two known temperatures by scribing on the glass stem at these temperatures. The remainder of the marks are distributed proportionally. For greater accuracy, more calibration points can be used. This covers the possibility of a nonuniform bore.

The greatest control for glass thermometers is obtained if the entire thermometer is immersed in the calibrating medium. If this is not possible, calibration can be done with partial or total immersion of the bulb.

Calibration of temperature-sensing devices is sometimes thought unnecessary. Situations can exist, however, that indicate that calibration should be carried out. Resistance thermometer stability is dependent on the element being free of residual strains, and thermocouple output can vary from tabulated values because of metal impurities and nonconsistencies and also due to aging of the metals.

Calibration is by one of two methods. The first is comparison with so-called primary standards. These include the ice point at 32° F, steam point at 212° F, oxygen point at –297° F, and so on. The second method uses secondary references, which include the freezing and boiling points of other elements and compounds.

Protection

In some applications, it is necessary to protect thermometer bulbs and thermocouple junctions from possible damage. This protection can be provided with either sockets or wells within which the thermometer bulb or thermocouple junction are located. It is important to remember that use of a socket or well reduces the response time for the measuring device. This reduction is due to two effects. The increased mass must be heated, which affects the temperature; and additional thermal resistance is introduced, which slows the response rate.

REVIEW MATERIALS

Important Terms

temperature	heat transfer
heat	conduction
degree	natural convection
scale	forced convection
BTU	linear coefficient of thermal expansion
calorie	Fahrenheit
volumetric coefficient of thermal expansion	Celsius
absolute scales	thermometer
absolute zero	bulb
Rankine	bimetallic thermometer
Kelvin	pressure-spring thermometer
phase	sublimation
thermistor	specific heat
thermocouple	thermal conductivity
pyrometer	convection
thermopile	radiation
thermal time constant	range
span	

Questions

1. What is the difference between temperature and heat?
2. What are the four chief scales for measuring temperature?
3. What are the three methods of transferring heat?
4. What is the difference between thermal expansion, thermal conductivity, and specific heat?
5. Why might the scales on a liquid-in-glass thermometer not be perfectly linear?
6. Which temperature-measuring devices may need compensation?
7. How does a bimetallic thermometer operate?
8. What are the four classes of pressure-spring temperature-measuring devices?
9. What is a thermistor?
10. Thermocouples are often used in measuring (high, medium, low) temperatures. Comment.
11. Why is it necessary to use a reference junction when making a temperature reading with a thermocouple?
12. What is a pyrometer, and how does it work?

13. What is a thermopile?
14. What is the difference between the range and span of a temperature-measuring device?
15. Why is the concept of a time constant used instead of the time required to reach the final temperatures?
16. If a relatively large abrupt change is taking place every 10 seconds, and the time constant of the temperature-measuring device is 10 seconds, how will the temperature readings be affected?
17. If the time constant in Question 16 is decreased to 1 second, how will the temperature readings be affected?
18. What thermometer might be a candidate for measuring the temperature at one point and indicating it at a point 400 feet away?
19. What factors affect the placement of a thermometer bulb?
20. What are the primary standards for temperature-measuring calibration?

Problems

1. Convert the following Fahrenheit temperatures to Celsius: 72°, −32°, 302°.
2. Convert the following Celsius temperatures to Fahrenheit: 190°, 60°, −10°.
3. Convert the following Fahrenheit temperatures to Rankine: 80°, 20°, −10°.
4. Convert the following Rankine temperatures to Fahrenheit: 100°, 400°, 800°.
5. Convert the following Celsius temperatures to Kelvin: −20°, 20°, 120°.
6. Convert the following Kelvin temperatures to Celsius: 100°, 20°, 300°.
7. Convert a temperature of 100° R to a Kelvin temperature.
8. At what temperature are the Celsius and Fahrenheit readings the same?
9. It is necessary to make a change in temperature of 30 Fahrenheit degrees. How many Celsius degrees would this be?
10. How much heat is required to raise the temperature of 2 pounds of copper from 50° F to 110° F?
11. How many degrees Fahrenheit will 100 BTUs raise the temperature of 10 pounds of mercury?
12. An object weighing 50 pounds is taken from a furnace at 900° F and put into 800 pounds of oil at 80° F. If the specific heat of the oil is 0.5 BTU/lb °F, what is the specific heat of the object? The final temperature is 100° F.
13. A rigid container contains 0.6 lb of gas. If the specific heat is 0.35 BTU/lb °F, what is the change in gas temperature if 100 BTUs of heat is added?
14. A thermal insulator has a cross-sectional area of 100 in² and is 1" thick. If the temperature difference between its two faces is 100° F, how much heat must flow through? The thermal conductivity is 0.0002 BTU/sec ft °F.

15. One end of a bar 18" long and 4 in² in cross-section is at 212° F, and the other end is at 32° F. If the thermal conductivity is 230 BTUs/hr ft °F, how much heat is flowing through the bar? Assume no heat flows from the sides of the bar.

16. A brick wall is 4" thick. What is the rate of heat transfer if one face is at 75° F and the other at 30° F? Assume k for the brick is 0.40 BTU/hr ft °F.

17. A wall is 1" thick. If its area is 10 ft² and its k is 0.2 BTU/hr ft °F, what is the cool-side temperature? The warm side is at 90° F and 1,000 BTUs/hr flows through it.

18. If equal-area slabs of steel and copper are to have the same amount of heat flow through them, which will be the thicker and by how much? The temperature change across each slab is 100° F.

19. A clock has a pendulum with an aluminum arm. How much does the length of the arm change if the temperature increases from 50° F to 80° F?

20. The length of a steel bridge is 1,000 feet. How much does its length change between –30° F and 90° F?

21. A steel tape measure of 100 feet is known to be the correct length at 70° F. If the temperature is 100° F on a day that a distance of 76.57 feet is measured, what is the correct distance?

22. Steel rails 60 feet long are laid end to end just contacting each other when the temperature is 100° F. How large will the gap between them be if the temperature drops to –30° F?

23. A wire is 10 feet long at 70° F. Its length increases by $1/16$" when heated to 170° F. What is the materials coefficient of linear expansion?

24. A copper rod 4" long at 70° F should reach a contact when the temperature reaches 350° F. What should the initial gap between the rod and the contact be?

25. A thermometer is calibrated over the interval from 200° F to 900° F. What are its span and range?

26. If a temperature-measuring system has a span of 1,000° F and an accuracy of ±1% of span, what is the possible error at a reading of 140° F?

27. Determine the time constant for a temperature-measuring device from which the following data were obtained:

Temperature, °F	Time, seconds
130	2
180	4
220	6
255	8
280	10
340	15
360	20
385	30
395	40
398	50

The original temperature of the device was 70° F, and the temperature of the media was 400° F.

28. In Problem 27, how much time would be required to reach 90% of the final value?
29. A thermometer has a time constant of 1 second. If the temperature of the media it is sensing suddenly increases from 100° F to 200° F, its readout after 2 seconds would be closer to which of the following: 182° F or 187° F?
30. For the following, plot the data and determine a time constant:

Time, seconds	Temperature, °F
0	20
5	93
10	138
15	162
20	183
30	195
40	201
50	203

Level

CHAPTER GOALS

After completing study of this chapter, you should be able to do the following:

- Understand several of the various level-measuring situations that exist.
- Explain the difference between continuous and single-point or position level measurement.
- Describe the concept of the dielectic constant.
- Work with the formulas involved with level measurement.
- Understand the working principles of both direct and indirect level-measurement devices.
- Know other considerations involved with the application of level-measurement devices.

Many industrial operations involve the use of liquids such as water, chemicals, fuel, and the like, or powder or a pelletized solid. These materials are often stored in containers such as tanks or hoppers to be used as needed. The surface level of the stored material is an indication of the amount remaining and hence is a possible application for measurement instrumentation.

BASIC CONSIDERATIONS

Basic Terms

The basic concept covered in this chapter concerns the level of material in a container. The material involved is most often a liquid; but, as will be noted, slurries and some forms of solids can be included. Another way of looking at the problem is to envision it as the sensing of interfaces. These interfaces could take the form of gas–liquid, gas–solid, liquid–liquid (where the two liquids are immiscible), and liquid–solid. From this viewpoint, most everyday situations fall into either the gas–liquid or gas–solid categories.

There are two general types of level measurement: *continuous* and *single-point* or *position*. Continuous level measurement gives the location of the level at any instant in time. Single-point measurement indicates when the level has reached the point under consideration. Continuous level measurement along with knowledge of the container geometry allows calculation of volume. From Figure 8-1, if the cross-sectional area of the container is one square foot, the volume contained changes by one cubic foot each time the level changes one foot. Single-point level measurement is useful for such situations as warning when overflow is imminent, alarm, and the like. As noted later, the volume-type measurements can also be used to calculate weight changes.

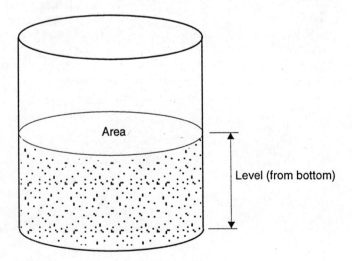

Figure 8-1 Knowing the level of the material and the cross-sectional area of the container allows calculation of the volume of the material.

Solids to be considered for level-measuring situations are in the form of powders, grains, or crystals. They must be essentially free flowing. To meet the free flowing condition, it is usually necessary that they be dry.

Later in the chapter, the term *dielectric constant* appears. A dielectric tends to be an insulating material. When placed between the plates of a capacitor, the capacitance increases. If the dielectric completely fills the space between the plates, the capacitance increases by a factor known as the dielectric constant. The dielectric constant value for air is 1; for pure water, it is 80. The higher the number, the more conductive the material.

Formulas Involved in Determining Level

Formulas used in making calculations related to level usually involve pressure in one way or another. Thus, it should be recalled that pressure increases with depth in a fluid according to:

$$\Delta p = \gamma \, \Delta h \qquad (8\text{-}1)$$

where Δp, the change in pressure, will be in lbs/ft² if the specific weight, γ, is in lbs/ft³ and Δh, the change in depth, is in feet.

Some level applications make use of the buoyancy principle. As described in Chapter 5, buoyancy is the upward force on an object that is partially or totally immersed in a fluid. It is a consequence of the way pressure is distributed around the object. In general, the buoyancy force can be calculated from the following equation:

$$B = \gamma \, (\text{displaced volume}) \qquad (8\text{-}2)$$

where the buoyant force, B, is in pounds if γ is in lbs/ft³, and the displaced volume is in ft³.

MEASURING DEVICES

There are two general classifications of devices for measuring levels. *Direct measuring devices* work with the level itself in providing a reading. *Indirect measuring devices* use something other than the level, such as a pressure or a pressure change.

Direct Level Measurement

Sight or Gage Glasses. Probably the simplest method for measuring a liquid level is with a sight or gage glass. As shown in Figure 8-2, the glass is mounted vertically alongside the tank, with the ends of the glass connected to the top and bottom of the tank. The liquid level in the glass must be the same as the level in the tank (if it were not, the liquid would move until it was).

In the application shown in Figure 8-2, the tank is closed. It would be possible for the pressure above the liquid level to be greater than atmospheric; that is, it could be a "pressurized tank." The sight glass would have to be capable of withstanding the pressure, but it would still give a correct level reading. Another application is shown in Figure 8-3, where the tank and glass are open to the atmosphere. The reason that the glass must be open to the atmosphere can be explained by first imagining that the tank and glass are empty. Then the tank filling process starts. If the top of the glass is closed, air will be trapped above the liquid. This air would have to be compressed as the liquid level increased. In such a situation,

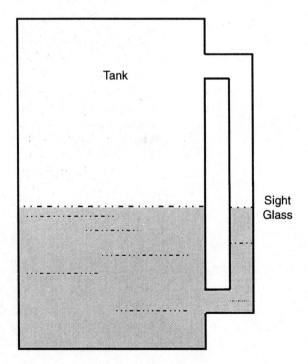

Figure 8-2 The level in the sight glass is the same as the level in the tank.

Figure 8-3 Both tank and sight glass are open to the atmosphere.

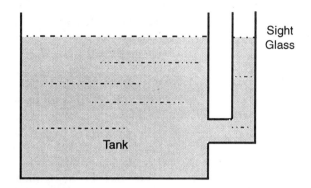

the levels of the liquid in the tank and sight glass would not be the same.

For the case of the open tank and glass, it would be possible to use a sight-glass liquid heavier than the liquid in the tank glass. In this way, the glass would not have to be as tall as the tank (Figure 8-4).

Limitations of sight or gage glasses include situations where the liquid discolors the glass to the extent that readings cannot be made or that cleaning would be required too frequently. Also, it is difficult to use sight or gage glasses if the liquid is very viscous or if it contains material that could settle out and clog the glass. Other limitations include the fact that tanks are often hard to get to and reading the glass might be difficult. Another limitation occurs if the working conditions are such that the glass could be broken. This could be hazardous to personnel, especially if the liquid were corrosive or had other undesirable properties.

Figure 8-4 Sight-glass liquid has higher specific weight than the liquid. *(From Jack W. Chaplin, Instrumentation and Automation for Manufacturing, © 1992, Delmar Publishers, Inc., Albany, N.Y.)*

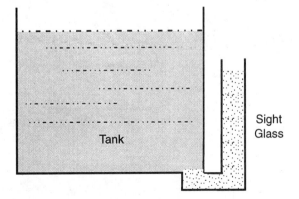

A sight or gage glass application that overcomes some of these limitations involves using a sight or gage "glass" not made from glass. If tubes are made from a stronger material, there are more possible applications. To overcome the fact that it is not possible to see through the tube, floats made from magnetic materials can be placed in the tubes and "followed" by indicating elements outside the tube.

Floats. After sight or gage glasses, the next simplest device involves a float. The float itself is "lighter" than the material whose level is being measured. The principle of buoyancy thus says that the float will ride up or down as the level changes. While there are many arrangements that can be used to convert the float level into a liquid level reading, one of most frequently used is shown schematically in Figure 8-5. The angle position device can be designed such that the correct liquid level is registered and recorded elsewhere. Other arrangements include float counterweights and connecting arm length-measuring devices. In general, errors introduced into the liquid level reading by the float itself can be minimized if the float is relatively large in surface area contacting the liquid and short in the direction of changing level. One of the everyday applications of the float method is in the fuel tank of an automobile.

Special float applications can be made for measuring the level of some solids. Sometimes called the *contact* method, Figure 8-6 shows that the amount of line necessary to reach the solid can be related to its level. It should be noted that the solid materials should be either somewhat free flowing or that some sort of agitation or vibration be used to ensure at least a partial evening of the solids' surface.

Figure 8-5 Level in tank is sensed by angle of connecting arm.

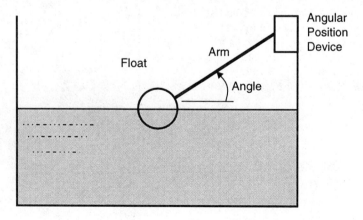

Figure 8-6 Levels of some bulk solids can be measured with the float method.

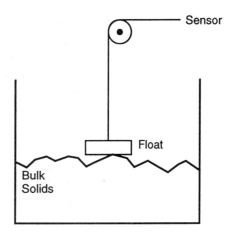

The advantages of the float method are that it is simple and that it is not seriously affected by changes in the density of the liquid. A disadvantage can occur if the liquid surface is wavy or otherwise disturbed. This can cause the float reading to vary erratically.

Displacers. Displacers are similar to floats except that their movement is usually severely restricted. Thus, it is the change in buoyant force on the object (Figure 8-7) that is related to the liquid level. In general, displacers should not be used when the liquid surface is greatly disturbed or when the liquid density undergoes significant changes. Displacers are usually more sensitive devices than floats.

Figure 8-7 Changes in liquid level cause changes in buoyant force on the displacer.

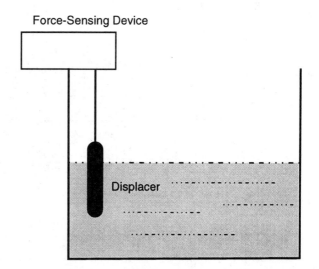

Probes. There are three general classifications of probes that can be used for measuring levels: Two are electrical in nature, and the third is ultrasonic. Depending on the type of probe, single-point or continuous level measurement can be made.

A *conductivity* probe consists of two electrodes placed in the tank (see Figure 8-8). When the liquid level rises enough so that both electrodes are in contact with the liquid, a conductive path is formed from one electrode to the other. When this occurs, a relay is energized, indicating that the level of interest has been reached. The liquid must be a conductor and should not be of such a nature to be dangerous if a spark should occur.

A *capacitance* probe (see Figure 8-9) usually consists of an inner rod and an outer shell. When the level of the liquid is below the probe, the air or vapor in the tank serves as the dielectric. When the liquid level reaches the probe, the liquid enters the space between the rod and shell. Because the dielectric constant of the liquid is much different than that of the air or vapor, a noticeable change in capacitance is observed. This method of level measurement is usually best applied when the liquid is essentially nonconductive. If the tank wall is made of metal, it can serve the same purpose as the outer shell; and only an inner rod is needed. This type of application could be used for continuous level measurement.

Ultrasonic probes use high-frequency sound waves to indicate level. One application consists of a transmitter, receiver, and gap in between (see Figure 8-10). An electric signal is supplied to the transmitter which converts the energy to sound

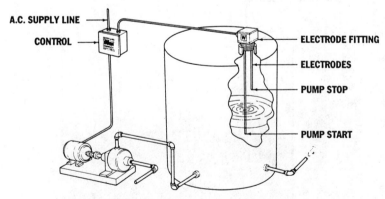

Figure 8-8 Conductive rod level measurement set-up and readout device. (Intrinsically safe revisions may be used where explosive gases are present.) *(Courtesy of Warrick Controls, Inc.)*

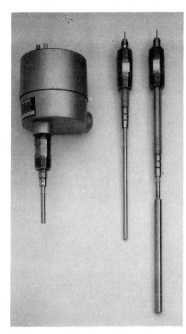

Figure 8-9 Capacitance probes and switch used for level measurement *(Courtesy of Princo Instruments, Inc.)*

Figure 8-10 Single-point ultrasonic probe indicates when gap is entirely beneath liquid surface.

waves. As soon as there is liquid in the gap, sound energy passes to the receiver; and the level is indicated. This type of application is obviously for single-point measurement. Another ultrasonic application is shown in Figure 8-11. As the level changes, the time required for the sound energy to pass from the transmitter to the level of the material being measured and back to the receiver changes. The time required for this cycle can then be related to the level. Note that continuous level measurement is possible. This type of ultrasonic application is desirable in some cases because it does not disturb the material in the tank. It is also convenient for level measurement of some solids levels.

Before leaving the discussion of probes, it should be pointed out that those devices generally considered to be single-point level indicators can be used to give "change in level" measurements if more than one probe is used. Figure 8-12 shows that three probes could be installed in a tank and then the time between passing any two levels could be interpreted as a level movement time.

Another fairly common application of more than one probe would be where three probes were used. One probe could give a reading when a low level was reached (and perhaps a pump should be turned on); a second could indicate an upper level (the pump could be shut off); and the third could give warning if a dangerously high level had been reached and an alarm sounded.

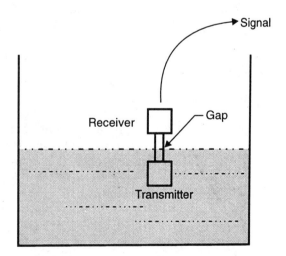

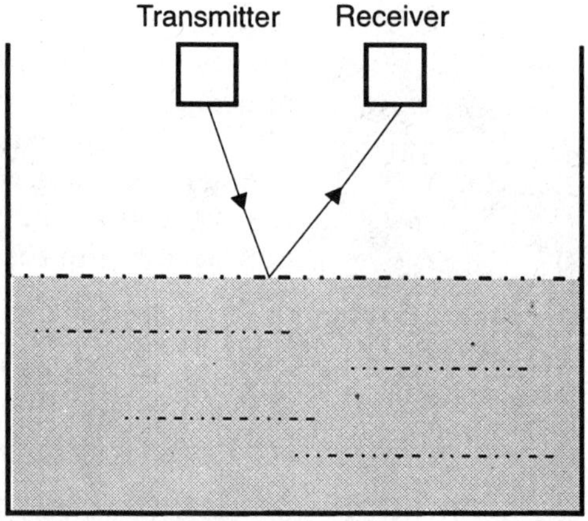

Figure 8-11 Continuous level ultrasonic probe arrangement.

Figure 8-12 Use of more than one probe can be used to approximate "change in level" time.

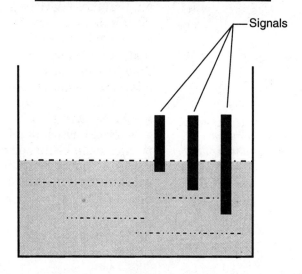

Indirect Level Measurement

Hydrostatic pressure is probably the most widely used indirect level measurement technique. Because the pressure at a given point in a liquid is directly related to the depth of the liquid above that point, measuring the pressure and knowing the liquid's specific weight allows calculation of the depth from Equation 8-1. This measurement could be made with a manometer as described earlier in this chapter (see Figure 8-4). Probably the most popular method, however, is the use of a pressure gage. It is possible to use a pressure gage mechanism and, knowing the liquid's specific weight, to make a gage face read directly in a length (e.g., inches, feet, or cm of liquid). It is also possible to use a pressure transducer and then convert the reading into a liquid level. The latter method lends itself well toward making a permanent record of the level. If the tank is closed in a manner similar to that shown in Figure 8-2, a differential pressure transducer can be used to measure the net pressure due to the liquid depth.

EXAMPLE 8-1

A pressure gage located at the base of an open tank registers 13.3 psi. If the liquid in the tank has a specific weight of 56 lbs/ft^3, what is the level of the fluid above the tank base?

From Equation 8-1:

$$h = p/\gamma = \left(13.3 \frac{lbs}{in^2}\right)\left(\frac{144 \ in^2}{ft^2}\right) / 56 \frac{lbs}{ft^3}$$

$$h = 34.2 \ ft$$

Bubbler Devices. The use of a bubbler system requires a source of clean air or gas. As shown in Figure 8-13, the air or gas is forced down the tube, whose lower end is open and placed near the bottom of the tank. Because the specific weight of the air or gas is so much less than that of the liquid in the tank and the air or gas is moving so slowly down the tube, the pressure gage near the top of the tube gives a reading very nearly that of the pressure at the bottom of the tube. Using the specific weight of the liquid, the depth or level of the fluid can be calculated. The reading must be made at the lowest pressure that will cause bubbles. Changes in the liquid level will require an increase or decrease in the pressure required, which will again indicate the liquid level. One of the advantages of the bubbler method is that if the liquid is corrosive, only the relatively inexpensive tube need be occasionally replaced.

EXAMPLE 8-2

Air just begins to bubble from the bottom of a vertical pipe submerged in water when the air pressure is 4.7 psi. How far above the bottom of pipe is the water level?

Figure 8-13 Level measurement using the bubbler technique.

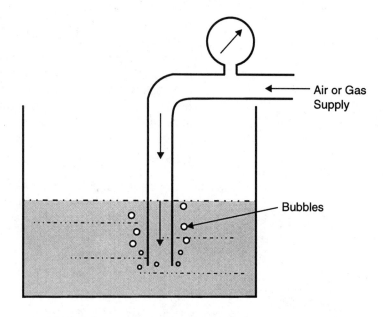

From Equation 8-1:

$$h = p/\gamma = 4.7 \frac{\text{lbs}}{\text{in}^2} \frac{144 \text{ in}^2}{\text{ft}^2} / 62.4 \frac{\text{lbs}}{\text{ft}^3}$$

$$h = 10.85 \text{ ft}$$

Radiation Devices. Level can also be measured using radioactive material. These devices are normally used when the liquid is corrosive, too hot, or otherwise not appropriate for installing equipment inside the tank. For single-point level measurement, such as the top location in Figure 8-14, only the top radiation source and the detector are needed. If further indication of level changes is desired, more sources can be added as shown. The principle of operation is that as the liquid level passes the source, the rays passing from that source to the detector are disturbed. Although this method can be used to solve difficult level-measurement problems, it is a relatively expensive method and requires the sometimes undesirable handling of the radioactive material.

Weight Method. It is possible to determine the level of material (liquid or solid) in a container by weighing the container and the material together and then subtracting the previously

Figure 8-14 Radiation level measuring set-up.

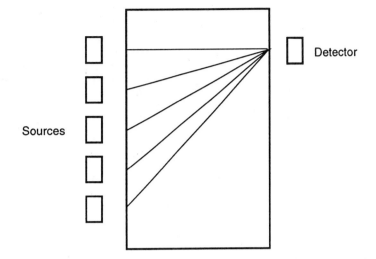

determined weight of the container (see Figure 8-15). Thereafter, any change in indicated total weight will be a result of the change in level of the material. This method is well suited for making a continuous record of level. It is necessary, however, to have a constant material-specific weight, particle size, and moisture content for this method to give accurate results. One of the obvious advantages is that the material need not be disturbed or come in contact with a sensor.

Figure 8-15 Load cell for weight method of measuring level *(Courtesy of Daytronic, Inc.)*

EXAMPLE 8-3

A container with inside dimensions of 12" and 36" contains a liquid with a specific weight of 70 lbs/ft³. What is the depth of the fluid in the container if it weighs 20 lbs and the total weight is 160 lbs?

$$\text{Weight of liquid} = 160 - 20 = 140 \text{ lbs}$$

$$\text{Volume of liquid} = \frac{140 \text{ lbs}}{70 \text{ lbs/ft}^3} = 2 \text{ ft}^3$$

$$\text{Cross-sectional area of container} = (1 \text{ ft})(3 \text{ ft}) = 3 \text{ ft}^2$$

$$\text{Depth of liquid} = \frac{2 \text{ ft}^3}{3 \text{ ft}^2} = 0.667 \text{ ft}$$

APPLICATION CONSIDERATIONS

Selection and application of a level measuring system can be affected by many considerations. The following discussion points out some of these items.

Floats. With regard to float shape, Figure 8-16 shows a cylinder. As the float sits in liquid, it displaces a volume of liquid which has a weight equal to its' weight. If the cylinder diameter is d, the buoyant force for a submersion of h_s is as follows:

$$\text{float weight} = \text{buoyant force} = \gamma_{liquid} \left(\frac{\pi d^2}{4}\right) h_s \qquad (8\text{-}3)$$

In Equation 8-3, if γ_{liquid} is given in lbs/ft³, d and h_s will have to be measured in feet or, alternatively, a correction factor applied to ensure correct values. Since the float weight is constant, the buoyant force is constant; and it can be seen from Equation 8-3 that γ_{liquid} and h_s are inversely related. As the liquid becomes "heavier," the depth of submersion decreases. In some instances, it may be desirable to give extra consideration to float shape, since some shape–submersion combinations are more stable than others and would tend to bounce or wobble less. As a general rule, a relatively flat float is prob-

Figure 8-16 Float shape is important in some level-measurement applications.

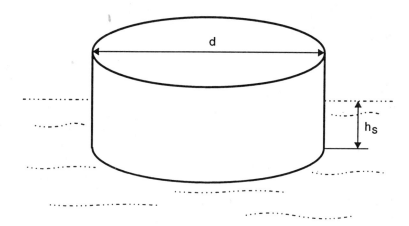

ably a good choice. Methods of making the float position known to an observer include a gear train and pointer connected to the float. Care must be taken if the tank is under pressure greater than atmosphere. Magnetic drives and push rods can be used in these cases.

EXAMPLE 8-4

A hollow float is in the form of a 6" cube. It weighs 2 lbs. How high on the float will water be?

Float weight = buoyant force = 2 lbs

$$2 \text{ lbs} = 62.4 \frac{\text{lbs}}{\text{ft}^3} (0.5 \text{ ft})(0.5 \text{ ft})(h_s)$$

$h_s = 0.1282$ ft, or 1.538 inches

Displacers

The buoyant force on a displacer (Figure 8-17) is given by the following equation:

$$\text{force on displacer} = \gamma_{\text{liquid}} \frac{\pi d^2 L}{4} \qquad (8\text{-}4)$$

where γ_{liquid} is the specific weight of the liquid, d is the float diameter, and L is length of the displacer submerged in the

Figure 8-17 Force on a displacer is dependent on depth of submersion.

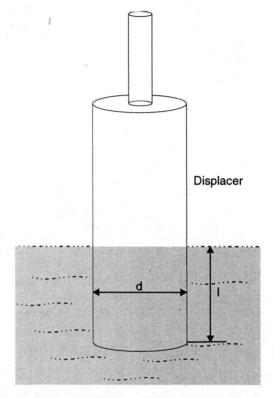

liquid. Care must be used so as to have compatible units. From this equation, it is possible to calculate the largest force to be expected. This can be done by replacing L with the maximum useable displacer length. Note that the liquid should not rise above the top of the displacer. If this should happen, increases in level will not be registered by the readout device.

EXAMPLE 8-5

A displacer is being used to measure the change in water level. How much will the level have changed if the force on the displacer changes 6.13 lbs? The displacer diameter is 6 inches.
From Equation 8-4:

$$6.13 \text{ lbs} = 62.4 \frac{\text{lbs}}{\text{ft}^3} \frac{\pi(0.5)^2 L}{4}$$

$$L = 0.5 \text{ ft, or 6 inches}$$

Capacitance Devices

Some capacitance applications require that the tank sides be parallel to the probe. Care should be taken to heed manufacturers' directions. Also, it may be necessary to check the liquid dielectric constant occasionally to see if it has changed. If so, adjustments are in order.

Pressure Gages

The use of pressure to infer a liquid level requires certain items of care with respect to the gages. These items include the following:

- Solids from the liquid may plug the line to the gage.
- The liquid should not corrode the gage; it may be necessary to use a sealing fluid as discussed in Chapter 5.
- The distance between the gage and the tank must be given consideration (see Chapter 5).

Bubbler Devices

Special precaution should be taken when installing a bubbler device to ensure a supply of air or gas. An undetected interruption in the air or gas supply obviously leads to erroneous level readings. Also, bubble movement should be started before the liquid is introduced to the tank so as to keep the liquid from entering the bubbler tube.

Radiation Devices

Users of radiation devices should familiarize all personnel to be associated with their operation with the possible hazards involved. This includes transportation, storage, and disposal. Overemphasis in this matter is hardly possible.

Other Liquid Level-Measurement Considerations

Liquid level measurement can be effected by frothing at the surface or turbulence in the liquid. Care should be taken to minimize their effects. Frothing can sometimes be controlled by adding a small amount of chemical which has no other effect. Baffles can reduce the effects of turbulence. The liquid level in a tank can also resonate or slosh around. If this should occur to an unusual degree, level measurement can be affected. Usually, resonance can be controlled by initial tank design or by proper use of baffles.

Solids Level Measurement

If the solids involved are in the form of powder, grains, or crystals that do not sufficiently self-level, it is possible to use a paddle wheel as shown in Figure 8-18. The wheel is driven by an electric motor. When the wheel is not in contact with the material in the tank, relatively little torque is required. When the level of the material increases sufficiently to touch the wheel, the torque begins to increase; and as more and more material enters the tank, the torque continues to increase. In many cases, this increase in torque can be related to the material level.

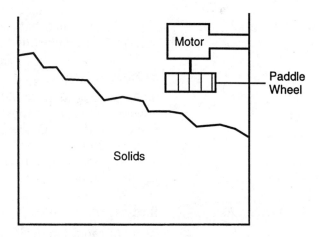

Figure 8-18 When the level of the solids reaches the paddle wheel, the torque to the wheel changes.

REVIEW MATERIALS

Important Terms

continuous level measurement
single-point or position measurement
dielectric constant
direct measuring
indirect measuring
sight glass
float
weight method

contact method
displacers
conductivity probe
capacitance probe
ultrasonic probe
bubbler device
radiation device

Questions

1. Discuss the difference between a direct and an indirect level-measuring device.
2. What is the principle that allows floats and displacers to be used as level-measuring devices? Explain the way it affects the measuring devices.
3. What does the term *dielectric* mean?
4. Which level-measuring device is good for application in the measurement of corrosive liquids?
5. If an open tank holds a liquid with a varying density, which of the following would be the best choice for measuring level and why?
    float   bubbler   pressure gage
6. How does a liquid's specific gravity or specific weight enter into level measurement?
7. Discuss the similarities and differences between a sight glass and a U-tube manometer.
8. What might be some restrictions on the use of a pressure gage for making liquid level measurements?
9. Some liquid level measurements could be made with an ordinary yardstick or meterstick. Name some of the restrictions to this type of measurement.
10. Why should the inside diameter of a sight glass be as large as possible?
11. If a float is replaced with one different from the original, what sort of considerations must be made?

Problems

1. A pressure gage indicates 10 psi at the bottom of an open water tank. What is the level of water above the gage?
2. If the highest level to be measured with a bubbler system is 30 feet of water, what pressure is required for the bubbler?
3. A liquid with a specific gravity of 0.85 is contained in an open tank 8 feet in diameter and 10 feet tall. What should the range be for a pressure gage to be placed at the base of the tank?
4. A tank 20 feet tall is mounted on the top of a platform 55 feet high. A pressure gage 10 feet up from the bottom of the platform reads 26 psi. What is the level of water in the tank measured from the base of the tank?
5. A displacer is 10 inches in diameter and 40 inches long. If it is half submerged in a liquid with a specific gravity of 0.9, what is the buoyant force acting in it?
6. What pressure would be caused by a liquid column 27 feet high if the specific weight of the liquid is 110 lbs/ft^3?
7. If a column of liquid 15 feet high causes a force of 140 lbs on a circular area with a 5" diameter, what is the specific gravity of the liquid?

8. A float measuring a liquid level can move through a vertical distance of 6 feet. If a change in the liquid's specific gravity causes the float to be 1 inch higher than it should be, how much error is introduced for a full-scale measurement? What would the error be if only half of the 6 feet were being used?
9. A bubbler system is to be used to measure the level of water in a container. If the level will vary between 5 and 35 feet, what is the minimum bubbler pressure required?
10. What is the specific weight of a liquid if a column 5 feet high gives a pressure of 14.6 psi?
11. If a water level is 54 inches, what is the pressure at the base?
12. Liquid level is being measured by change in weight. If the tank itself weighs 70 lbs, what would be the water level above the tank bottom if the total weight of tank and water were 563 lbs? The inside dimensions of the tank are 23" x 11".
13. The level of a liquid with a specific gravity of 1.2 is being measured by a weighing system. If the tank cross-section is circular with an inside diameter of 2 feet, what would be the weight change for each change in level of 1 inch?
14. What will be the maximum specific gravity for a liquid whose level is to be measured by a bubbler system with 20 psi available for the bubbler? It is desired to measure a 16 foot depth of the liquid.
15. What will be the buoyant force range for a 4 inch diameter displacer 60 inches long when its submergence will move between 10 inches and 50 inches? The liquid is water.

Flow

CHAPTER GOALS

After completing study of this chapter, you should be able to do the following:

Understand the concepts of fluid velocity, static and dynamic pressure, laminar and turbulent flow, Reynolds number, and viscosity.

Make calculations based on energy considerations using the Bernoulli equation.

Describe the difference between flow rate and total flow and make calculations related to both.

Understand the continuity equation and make calculations.

Describe the source of pressure losses related to fluid flow and make calculations.

Explain the way orifice plates, Venturi tubes, and flow nozzles are used and the advantages and disadvantages of each.

Understand the operation of specific total flow and flow rate meters.

Describe what open channel flow is and the devices that can be used for its measurement.

Apply what you learn to selecting, installing, and calibrating flow-measuring devices.

Flow measurement in industry is important for several reasons. In some applications, it is necessary to blend or mix definite proportions of liquids. Optimum performance of some machines or operations requires a specific fluid flow rate. The cost of many fluids is based on measurement of the amount received through a flow line. Whether dealing with gases or liquids, it is often necessary to measure accurately the amount of fluid flowing.

BASIC CONSIDERATIONS

Basic Terms

There are several concepts and definitions that must be discussed before proceeding to the formulas involved in making calculations related to flow and to description of devices for measuring flow. Some of these concepts and definitions make use of topics covered in Chapter 5.

Velocity. Velocity is a measure of the speed and direction of movement of a fluid particle or group of particles. In most cases, it is speed that is of primary interest. The units used are typically feet per second (fps), feet per minute (fpm), or meters per second (mps). As noted in Chapter 5, pressures associated with flowing fluid are either *static* (in which the effect of fluid velocity is not felt) or *dynamic* (including the effect of "stopping" the fluid).

Flow Patterns. If the average fluid velocity is relatively slow, the fluid particles move in a smooth fashion and tend to stay in layers. This layer-like movement is called *laminar* flow. As the arrows in Figure 9-1 show, the fluid at the center travels the fastest, while the particles in contact with the wall do not move. This type of distribution of velocity is referred to as *parabolic.* If, however, the average fluid velocity is relatively fast, the particles also tend to have movement perpendicular to the overall direction of flow. The result is a "mixed-up" flow, which is called *turbulent.* Figure 9-2 shows the average velocities across the cross-section of the flow area. Note that the parabolic shape has been flattened and that there is a sharp drop in velocity near the walls. In either laminar or turbulent flow, the region of more slowly moving fluid near the wall is called the *boundary layer.* It was observed by Osborne Reynolds in the 1880s that the change from laminar to turbulent flow could be predicted if certain quantities were known. These quantities relate the density and viscosity of the fluid to a dimension describing a length measurement, such as width or diameter, and the average fluid velocity. A combination of these quantities in a certain form is known as the *Reynolds number.* It is generally accepted that if the Reynolds number (R) for flow in a pipe is equal to 2,000 or less, the flow will definitely be laminar. From 2,000 up to a not-so-well-defined upper number (sometimes 5,000 is used), the flow could be either laminar or turbulent, depending on other factors. This region is referred to as the *intermediate* Reynolds number region. Beyond the upper limit of the intermediate region, the flow is always turbulent.

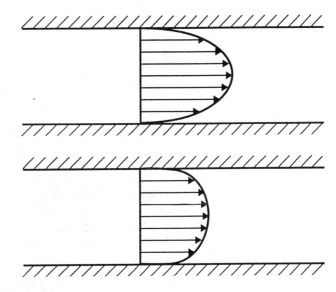

Figure 9-1 In laminar flow, the fluid travels in layers. The fastest moving fluid is at the center.

Figure 9-2 In turbulent flow, the average velocities away from the wall tend to be more uniform than in laminar flow.

The term *viscosity* refers to the resistance of a fluid to being flowed. It is a property of the fluid and may be thought of as the friction occurring as the fluid particles move past one another. Syrup has a higher viscosity than water, which, in turn, has a higher viscosity than gasoline. There are several different units used for measuring viscosity, including poise, centipoise, stokes, centistokes, and SAE numbers, among others. See Table 9-1 for viscosity conversions. In general, the viscosity of a liquid decreases as its temperature increases, while the viscosity of a gas increases with temperature increases.

Table 9-1 Conversion Factors for Dynamic and Kinematic Viscosities

Dynamic Viscosities		Kinematic Viscosities	
1 lb sec/ft^2	= 47.9 Pa s	1 ft^2/sec	= 9.29 x 10^{-2} m^2/sec
1 poise	= 10 Pa s	1 stoke	= 10^{-4} m^2/sec
1 centipoise	= 2.09 x 10^{-5} lb sec/ft^2	1 m^2/sec	= 10.76 ft^2/sec
1 poise	= 100 centipoise	1 stoke	= 1.076 x 10^{-3} ft^2/sec

Energy Factors. Many flow calculations are based on energy considerations. The most basic of these considerations concerns the relationships between fluid velocity, pressure, and the height of the fluid above a given reference point. The name most frequently associated with this relationship is the *Bernoulli* equation; and in its basic form, this equation states that energy is conserved. This means that the total energy of the fluid at one point in the flow must be equal to the total energy at another point. Other forms of the equation allow for such items as energy loss through friction and the addition of energy to the flow with a pump.

Pressure losses in a flowing fluid result from two general causes. The first of these is friction, either within the fluid itself or between the fluid and the boundaries within which the fluid is flowing. The second cause relates to the fluid impacting on an object, such as the wind blowing against the wall of a building. In some flow cases, it is necessary to take these losses into account when making calculations.

The laws governing fluid flow hold whether the fluid is a liquid or a gas. Many of the equations and formulas can be used directly on either. In other cases, however, the fact that the gas is compressible makes the equations or formulas more complicated and the calculations more difficult. Care must be taken to use the appropriate equations and formulas.

Types of Flow. There are two general types of flow measurement. *Flow rate* refers to the amount of fluid passing a given point in any given interval of time. Typical units are gallons per minute (gpm), cubic feet per minute (cfm), cubic feet per second (cfs), liters per minute, and the like. *Total flow* is the amount of flow past a given point over some length of time. Its units are gallons, cubic feet, liters, and the like. See Table 9-2 for flow rate conversion factors. Occasionally, total flow may be given in terms of the weight, that is, pounds.

Table 9-2 Flow Rate Conversion Factors

1 gallon/minute	=	6.309×10^{-5} m^3/sec
1 liter/minute	=	16.67×10^{-6} m^3/sec
1 gallon/minute	=	3.78 liters/minute
1 ft^3/sec	=	449 gallons/minute
1 gallon/minute	=	0.1337 ft^3/min
1 gallon/minute	=	0.00223 ft^3/sec

Formulas Used in Flow Calculations

Continuity. The *continuity* equation is one of the most basic equations in flow calculations. It states that if the overall flow rate in the system is not changing with time, the flow rate past any section of the flow system must be constant. In its simplest form:

$$Q = VA \qquad (9\text{-}1)$$

where

Q = flow rate

V = average fluid velocity at the section being considered

A = cross-sectional area of flow section being considered

The units on both sides of the equation must be compatible. One typical set is as follows:

$$\frac{ft^3}{sec} \text{ or cfs} = \left(\frac{ft}{sec}\right)(ft^2)$$

EXAMPLE 9-1

Fluid flows at a steady rate through a 6 inch diameter pipe with an average velocity of 3 feet per second. What is the flow rate?

$$Q = \left(\frac{3 \text{ ft}}{\text{sec}}\right)\left(\frac{\pi (0.5)^2 \text{ ft}^2}{4}\right) = 0.589 \text{ cfs}$$

Because 1 cfs = 449 gpm, 0.589 (449) = 264 gallons per minute are flowing.

Figure 9-3 shows a section of a flow system where there are two different areas, A_1 and A_2. From the continuity equation:

$$Q = V_1 A_1 = V_2 A_2 \qquad (9\text{-}2)$$

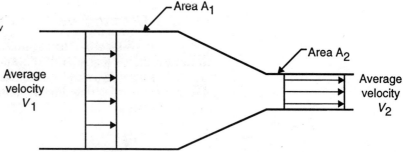

Figure 9-3 The continuity equation for steady liquid flow says that $V_1A_1 = V_2A_2$.

If Q, the number of cubic feet per second flowing through the system, is not changing with time, and area A_2 is smaller than area A_1, then velocity V_2 must be larger than velocity V_1.

EXAMPLE 9-2

If the fluid in Example 9-1 passes from the 6 inch pipe into a 6 inch square conduit, what is the average fluid velocity in the conduit? What is the flow rate in the conduit?

$$Q = V_1A_1 = V_2A_2$$

Since the flow is steady, the flow rate remains 0.589 cfs.

$$V_2 = \frac{0.589 \text{ ft}^3/\text{sec}}{(0.5)^2 \text{ ft}^2 \text{ sec}} = 2.36 \text{ ft/sec}$$

As it stands, Equation 9-2 is for a liquid. Although it holds for some gas flows, a more appropriate continuity equation for a gas is as follows:

$$\gamma_1 V_1 A_1 = \gamma_2 V_2 A_2 \qquad (9\text{-}3)$$

where γ_1 and γ_2 are the specific weights of the gas at the two sections. Typical units for Equation 9-3 become

$$\left(\frac{\text{lbs}}{\text{ft}^3}\right)\left(\frac{\text{ft}}{\text{sec}}\right)(\text{ft}^2) = \frac{\text{lbs}}{\text{sec}}$$

and the result is a weight rate of flow. Note that Equation 9-2 could be changed to a weight rate of flow by multiplying by γ, the specific weight.

Bernoulli Equation. In a given flow system, there is a relationship between pressure, fluid velocity, and elevation at any two points. Credit for defining this relationship is usually given to Bernoulli, an eighteenth-century scientist and mathematician. Considering points A and B in Figure 9-4, the Bernoulli equation is given as follows:

$$\frac{p_A}{\gamma_A} + \frac{V_A^2}{2g} + Z_A = \frac{p_B}{\gamma_B} + \frac{V_B^2}{2g} + Z_B \qquad (9\text{-}4)$$

where

p_A and p_B = pressures

γ_A and γ_B = specific weights

V_A and V_B = average fluid velocities

g = acceleration of gravity

Z_A and Z_B = elevations above a given reference level

Using typical units, each term in Equation 9-4 has the units of length as follows:

$$\frac{p_A}{\gamma_A} \longrightarrow \frac{\left[\frac{lbs}{ft^2}\right]}{\left[\frac{lbs}{ft^3}\right]} = [ft] \qquad (9\text{-}5)$$

$$\frac{V_A^2}{2g} \longrightarrow \frac{\left[\frac{ft^2}{sec^2}\right]}{\left[\frac{ft}{sec^2}\right]} = [ft] \qquad (9\text{-}6)$$

$$Z_A \longrightarrow [ft] = [ft] \qquad (9\text{-}7)$$

It is the Bernoulli equation from which the term *head* originated. Equation 9-4 can also be thought of as a conservation of energy equation with no energy loss between points A and B.

Figure 9-4 The Bernoulli equation relates the pressure, velocity, and elevation at points in the fluid, such as A and B.

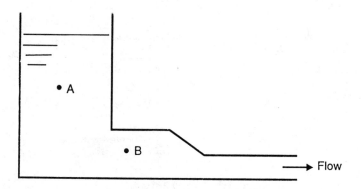

Thus, the first term represents energy stored as pressure; the second term represents kinetic energy (energy due to motion); and the third term represents potential energy or energy due to position. Equation 9-4 can be applied between any two points in the flow system. It should be noted the pressures used must be absolute pressures.

EXAMPLE 9-3

Using the system shown in Figure 9-5, calculate the velocity at point 3 and the velocity and pressure at point 2 if water is the system fluid. Water at point 1 on the surface of the reservoir is at atmospheric pressure, has zero velocity, and an elevation "h" above the reference line.

Applying Equation 9-4 to points 1 and 3:

$$\frac{14.7(144)}{62.4} + 0 + h = \frac{14.7(144)}{62.4} + \frac{V_3^2}{2(32.2)} + 0 \qquad (9\text{-}8)$$

From Equation 9-8, it can be seen that

$$V_3 = \sqrt{2(32.2)h} \qquad (9\text{-}9)$$

which states that the water velocity exiting is directly proportional to the square root of the difference in elevation between points 1 and 3.

The reader may be confused by the pressure at point 3 being equal to atmospheric pressure. Initial reasoning might indicate that because the fluid is moving at point 3, there is some pressure there above atmospheric. Recall, however, that there are two classes of pressure in flow situations: static and dynamic. In

Figure 9-5 Application of the Bernoulli equation to a flow system (Example 9-3).

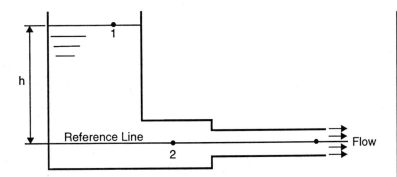

Equation 9-4, the pressures are static pressures. The fluid at point 3 indeed has a dynamic pressure but no static pressure above atmospheric.

Returning to Figure 9-5 and applying the Bernoulli equation between points 2 and 3:

$$\frac{P_2}{62.4} + \frac{V_2^2}{2(32.2)} + 0 = \frac{14.7(144)}{62.4} + \frac{V_3^2}{2(32.2)} + 0 \quad (9\text{-}10)$$

Because V_3 can be calculated from Equation 9-9 using a known value of "h", Equation 9-10 becomes an equation with only the pressure and velocity at point 2 unknown. From the continuity equation, it is known that

$$V_2 A_2 = V_3 A_3 \quad (9\text{-}11)$$

Normally the areas A_2 and A_3 are known, and V_2 can be calculated from:

$$V_2 = \left(\frac{A_3}{A_2}\right) V_3 \quad (9\text{-}12)$$

Using this value of V_2 in Equation 9-10 allows calculation of the pressure at point 2.

Using the following input information, typical values of the velocities at points 2 and 3 and the pressure at point 2 are as follows:

$$h = 10 \text{ ft}; A_2 = 0.25 \text{ ft}^2; A_3 = 0.125 \text{ ft}^2$$

$$V_3 = \sqrt{2(32.2)10} = 25.4 \text{ ft/sec} \quad (9\text{-}13)$$

$$V_2 = \frac{0.125}{0.25}(25.4) = 12.7 \text{ ft/sec} \qquad (9\text{-}14)$$

$$\frac{p_2}{62.4} + \frac{(12.7)^2}{2(32.2)} + 0 = \frac{14.7(144)}{62.4} + \frac{(25.4)^2}{2(32.2)} + 0 \qquad (9\text{-}15)$$

$$p_2 = 2{,}583 \text{ lbs/ft}^2 = 17.94 \text{ psia} = 3.24 \text{ psig}$$

The continuity equation could now be applied at point 3 such that

$$Q = V_3 A_3$$

In practice, the application is sometimes not so straightforward because of possible variations in the end of the conduit at point 3. The application takes the form:

$$Q = C_D \times VA \qquad (9\text{-}16)$$

where C_D is a discharge coefficient, which is dependent on the geometry involved. Before applying Equation 9-16, the reader should consult a flow handbook for discharge coefficient values.

Pressure Losses. The Bernoulli equation, as in Equation 9-4 does not take pressure losses into account. These losses fall into two general categories: (1) those associated with friction within the fluid and between the fluid and the walls containing it and (2) those associated with various fittings, such as elbows, tees, valves, and the like. Losses resulting directly from friction can be estimated using an equation of the following form:

$$h_L = f \frac{L}{D}\left(\frac{V^2}{2g}\right) \qquad (9\text{-}17)$$

where

h_L = head loss

f = friction factor

L = length of pipe under consideration

D = diameter of pipe under consideration

V = average fluid velocity in section under consideration

g = acceleration of gravity

The friction factor f depends on the Reynolds number for the flow and the roughness of the walls within which the fluid is flowing. Further information is available in flow handbooks.

EXAMPLE 9-4

Water flows through a pipe with an inside diameter of 1 inch. If the average velocity in the pipe is 8 feet per second, what is the head loss in 100 feet of pipe? Assume a friction factor of 0.02.

$$h_L = 0.02 \left(\frac{100 \text{ ft}}{1/12 \text{ ft}}\right) \frac{(8 \text{ ft/sec}^2)}{2\,(32.2 \text{ ft/sec}^2)} = 23.9 \text{ ft}$$

Note that this would be equivalent to

23.9 ft (62.4 lbs/ft^3) = 1,491 lbs/ft^2, or 10.37 psi

Flow losses associated with the various fittings are sometimes called *minor* losses. If the length of the flow system is rather long, these losses are indeed small when compared to the friction losses. If, however, the flow system length is rather short, the minor losses can become relatively large. The usual method of estimating minor losses is with an equation of the following form:

$$h_L = \frac{KV^2}{2g} \qquad (9\text{-}18)$$

where

h_L = head loss due to minor loss

K = head loss coefficient

Table 9-3 Typical Head Loss Coefficient Factors for Fittings

Threaded ell—1"	1.5
Flanged ell—1"	0.43
Threaded tee—1" inline	0.9
branch	1.8
Globe valve (threaded)—1"	8.5
Gage valve (threaded)—1"	0.22
Coupling or union—1"	0.08
Bell-mouth reducer	0.05

V = average fluid velocity through fitting under consideration

g = acceleration of gravity

Values of K can be found in flow handbooks and are usually dependent on items such as pipe size, Reynolds number, and others. See Table 9-3 for typical values.

EXAMPLE 9-5

Calculate the head loss through a combination of the following fittings. The average fluid velocity is 6 feet per second.

	Fittings	K
4	90° ells	1.5
2	tees	0.8
1	gate valve	0.22
10	couplings	0.085

$$h_L = \left[4(1.5) + 2(0.8) + (1)(0.22) + 10(0.085) \right] \frac{6^2}{2(32.2)}$$

$$= (6 + 1.6 + 0.22 + 0.85)(0.559)$$

$$= 4.85 \text{ ft}$$

After losses due to friction and fittings has been estimated, it is possible to use another form of Equation 9-4:

$$\frac{p_A}{\gamma_A} + \frac{V_A^2}{2g} + Z_A = \frac{p_B}{\gamma_B} + \frac{V_B^2}{2g} + Z_B + h_{L\text{friction}} + h_{L\text{fittings}} \quad (9\text{-}19)$$

This equation can be interpreted as showing that the energy content of the fluid at point A is equal to the energy content at point B *plus* any losses between the two points.

Form Drag. When a flowing fluid impacts on an object, a force is exerted on the object resulting from the stopping of the fluid or the changing of its path. Thus, the wind exerts a force on a sign or on the wall of a building. Any device protruding into or across a pipe cross-section will be subjected to a force when a fluid moves through the pipe. The force resulting from these examples is called *form drag*, implying that the force depends on the "form" of the object. It can be estimated from the following equation:

$$F = C_D \left(\frac{\gamma}{g}\right) A V^2 \tag{9-20}$$

where

F = force on the object

C_D = drag coefficient

γ = fluid specific weight

g = acceleration of gravity

A = cross-section of the object "seen" by the flow

V = average velocity of the fluid

As an example of the cross-section seen by the fluid, air moving past a ball "sees" a circular area calculated using the radius of the ball. Wind blowing perpendicular to a pipe would "see" a rectangular area equal to the length of the pipe multiplied by its diameter. The drag coefficients for various objects are available in flow handbooks and are usually related to a type of Reynolds number, one which involves some length measurement of the object. See Table 9-4 for typical values.

EXAMPLE 9-6

What is the force acting on a 3 inch diameter ball traveling through the air at 130 feet per second? Assume that the air specific weight is 0.0765 lb/ft³ and C_D = 0.5.

$$F = 0.5 \left(\frac{0.0765 \text{ lb/ft}^3}{32.2 \text{ ft/sec}^2}\right) \left(\frac{\pi (0.25)^2 \text{ ft}^2}{4}\right) \left[(130)^2 \frac{\text{ft}^2}{\text{sec}^2}\right]$$

$$= 0.985 \text{ lb}$$

Table 9-4 Typical Drag Coefficient Values for Objects Immersed in Flowing Fluid

Circular cylinder with axis perpendicular to flow	0.33 to 1.2
Circular cylinder with axis parallel to flow	0.85 to 1.12
Circular disk facing flow	1.12
Flat plate facing flow	1.9
Sphere	0.1+

MEASURING DEVICES

There are many ways of measuring the flow of liquids and gases. The type of material flowing, its temperature, viscosity, vapor pressure, conductivity, and amount of suspended material all have an influence on the possible method or methods to be used. The general classifications to be considered are *flow rate meters, total flow meters,* and *mass flow meters.*

Flow Rate

Differential Pressure. One of the most commonly used methods for determining flow rate is by relating the flow rate to the drop in pressure across a restriction. The higher the flow rate through a given restriction, the greater the pressure difference. Three widely used differential meters are the *orifice plate, Venturi tube,* and *flow nozzle.*

The orifice plate device is the simplest of the three. It consists of a plate, usually made of metal, with a hole through which the fluids flows. Three commonly used orifice plates are shown in Figures 9-6, 9-7, and 9-8. The hole in the center of the *concentric* plate coincides with the center of the pipe through which the fluid flows. The circular hole in the *eccentric* plate and the partially circular hole in the *segmental* plate are located as shown to allow suspended solids to pass through. The plates usually have sharp edges as shown in Figure 9-9 with the downstream edge beveled so as to encourage a clean-cut separation of flow and a precise, repeatable flow pattern. Orifice plates can be used in flow with small amounts of dissolved air if small vent holes are placed in the plates near the top.

Orifice plates are usually held in place between flanges (Figure 9-10). With proper flange and plate design, the plate can be rather easily replaced. To measure the difference in pressure upstream and downstream from the orifice plate, pres-

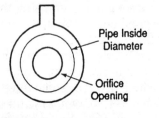

Figure 9-6 Concentric orifice plate *(From Jack W. Chaplin, Instrumentation and Automation for Manufacturing, © 1992, Delmar Publishers, Inc., Albany, N.Y.)*

Figure 9-7 Eccentric orifice plate *(From Jack W. Chaplin, Instrumentation and Automation for Manufacturing, © 1992, Delmar Publishers, Inc., Albany, N.Y.)*

Figure 9-8 (*below*) Segmental orifice plate *(From Jack W. Chaplin,* Instrumentation and Automation for Manufacturing, *© 1992, Delmar Publishers, Inc., Albany, N.Y.)*

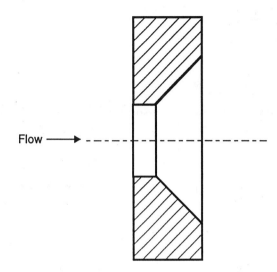

Figure 9-9 (*right*) Cross-section of concentric orifice plate.

sure *taps* or holes are placed at specific locations. Three commonly used tap locations are *flange, pipe,* and *vena contracta.* Flange taps are located in the bottom of the flanges holding the orifice plate in place. Pipe taps are located in the pipe upstream and downstream from the restriction. The term *vena contracta* refers to the shape of the fluid boundary as it passes through the orifice. Vena contracta taps are located at specific distances with respect to the orifice plate.

A Venturi tube is a restriction to flow with a specified reduction and following increase in flow area (Figure 9-11). Venturi tube application is usually for larger pipe sizes and can be used for flows with a relatively large amount of suspended particles. Pressure taps for a Venturi tube are located at the maximum and minimum tube diameters.

The flow nozzle represents a compromise between an orifice plate and a Venturi tube (Figure 9-12). It resembles the front half of a Venturi tube. Location of pressure taps is specified by the manufacturer and is critical to proper use of the nozzle.

Application of orifice plates, Venturi tubes, and flow nozzles involves consideration of the *beta ratio*, that is, the ratio of the restriction diameter to the pipe diameter. It is frequently given as d/D. A compromise is usually reached between small beta ratios with correspondingly high pressure (and energy)

Figure 9-10 Orifice plates are usually held in place by flanges.

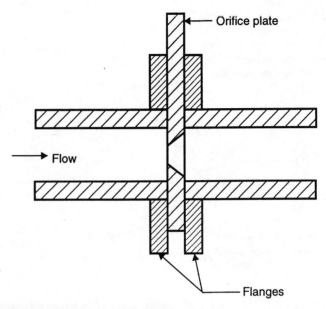

losses and large beta ratios which may not give sufficiently large pressure drops to allow accurate flow measurement. Typically, the ratio is between 0.2 and 0.6.

The orifice plate is the simplest of the three devices and the least expensive. It is the easiest to install or replace. Pressure (energy) losses are highest for orifice plates, and the plates are the most subject to damage or erosion. The Venturi tube is the most expensive and, properly used, the most accurate. It can be used with large beta ratios and has the least pressure (energy) loss. In addition, it wears well and is not significantly affected by sediment in the flow stream. Replacement of a Venturi tube is generally much more time consuming than for an orifice plate. The flow nozzle falls between the Venturi

Figure 9-11 Cross-section of Venturi tube. Note the decrease in static pressure at the throat (smallest part) and subsequent increase downstream.

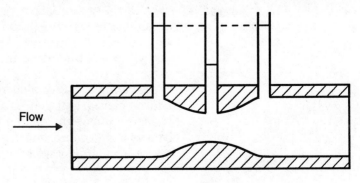

Figure 9-12 Flow nozzle *(From Jack W. Chaplin,* Instrumentation and Automation for Manufacturing, *© 1992, Delmar Publishers, Inc., Albany, N.Y.)*

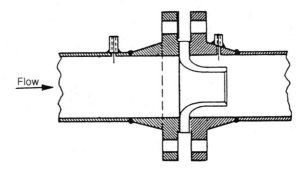

tube and the orifice plate with regard to cost, replacement time, pressure loss, and wear. It truly offers a compromise between them.

A change in flow direction as in an elbow (Figure 9-13) is accompanied with a pressure difference between the inside and outside of the curve. This pressure difference can be used to measure flow rate. The elbow meter is unlikely to clog and is relatively wear resistant. It is possible that the pressure taps could clog; thus, their placement should be carefully considered.

A Pitot-static tube can be used to measure flow rate (Figure 9-14). The difference between the impact or dynamic pressure (from the tap facing the flow) and the static pressure can be related to the flow rate. Alignment of the tube to the oncoming flow must be done carefully. In reality, the Pitot-static tube indicates a fluid velocity; and thus the velocity must be associated with a cross-sectional area in order to obtain a flow rate. Applying the Pitot-static tube requires either moving the tube across the flow cross-section to establish an av-

Figure 9-13 Elbow flow element *(From Jack W. Chaplin,* Instrumentation and Automation for Manufacturing, *© 1992, Delmar Publishers, Inc., Albany, N.Y.)*

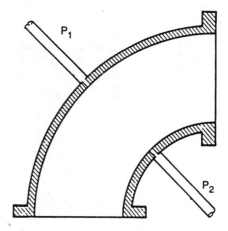

erage velocity for the section or calibrating the tube for one specific flow area. Two disadvantages associated with Pitot-static tubes are that they can become plugged with sediment and that the pressure difference sensed may not be large enough to give the desired accuracy for the flow rate under consideration.

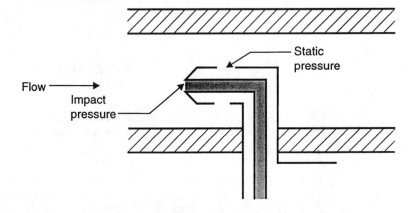

Figure 9-14 Pitot-static tube for measuring fluid velocity.

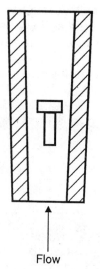

Figure 9-15 A rotameter consists of a tapered tube (usually glass) and a float.

Variable-Area Meters. *Rotameters* are one of the most widely used flow rate–measuring devices. They consist of a vertical tapered tube with the flow moving upward (Figure 9-15). Inside the tube is a *float* from which the readings are made. Flow past the float causes it to move to an equilibrium location such that its buoyed weight is balanced by the force resulting from the pressure difference between the top and bottom of the float. The higher the flow rate, the larger the cross-section of the tube must be in order that the force resulting from the pressure difference force just match the float's buoyed weight. In contrast to orifice plates, Venturi tubes, and flow nozzles, which have constant areas and pressure differences that change with flow rate, the rotameter has a variable area which leads to a constant pressure drop or force on the float. Rotameters can be used for measuring flow rates of both gases and liquids. Tube material includes glass, plastic, and metal. If the tube material or fluid color is such that the float is not visible, a magnetic following device on the outside of the tube can be used to show the float location. The upper limit of the flow measuring capacity of a rotameter is relatively low compared to other devices. To overcome this limitation, rotameters are used in combination with an ori-

Figure 9-16 Bypass flow meter *(Courtesy of Wallace & Tiernan, Inc.)*

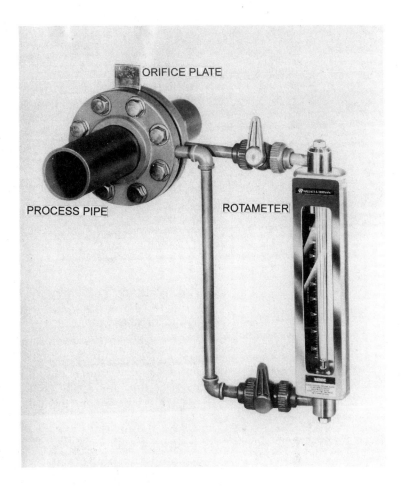

fice plate (Figure 9-16). With this set-up, the pressure drop between the flanges causes a flow through the rotameter which can then be related to the main flow rate.

Other Flow Rate Meters. A *turbine* flow meter usually consists of a section of pipe, a rotor with blades mounted in the center of pipe and a sensor mounted outside the pipe (Figure 9-17). The bladed rotor (turbine wheel) turns as the fluid passes through. Each rotation of the rotor is sensed, usually with an electrical-magnetic device. In general, the rate of rotor rotation is directly related to the flow rate; thus, the flow rate can be obtained by counting the number of rotations in a given time period. Turbine flow meters are generally accu-

Figure 9-17 Turbine flow meter cross-section *(From Jack W. Chaplin,* Instrumentation and Automation for Manufacturing, *© 1992, Delmar Publishers, Inc., Albany, N.Y.)*

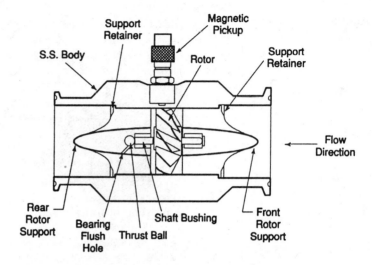

rate and have good ranges. They are relatively expensive and operate best with clean fluids. They can also be affected by fluid viscosities.

An *electromagnetic* flow meter usually consists of a pipe made of nonconducting material with two electrodes mounted opposite each other on the pipe wall (Figure 9-18). The inside ends of the electrodes contact the fluid (which must be an electrical conductor) flowing. A magnet encircles the pipe such

Figure 9-18 Electromagnetic flow meter configuration *(From Jack W. Chaplin,* Instrumentation and Automation for Manufacturing, *© 1992, Delmar Publishers, Inc., Albany, N.Y.)*

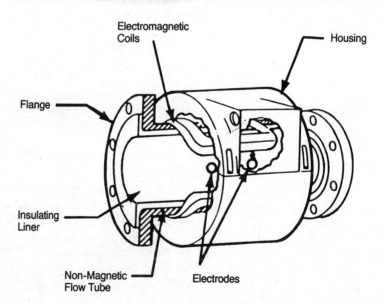

that its field is perpendicular to the electrodes. As the fluid moves through the pipe and magnetic field, a voltage is generated that depends on the strength of the magnetic field, the distance between the electrodes, and the velocity of the fluid. With proper design, the voltage can be used to indicate flow rate. Advantages include good accuracy, essentially no pressure loss, and no other restrictions to flow. Also, fluid viscosity, fluid density variations, and suspended material have very little effect on operation. Care must be taken to keep the electrodes from acquiring a coating that would affect proper operation. These meters are usually relatively expensive.

The general term *vortex* is used to denote a swirling or rotating type of fluid motion. Two different types of vortex motion are used to measure flow rate. The first of these is used with gases and is called a *vortex precession* meter (Figure 9-19). The meter body contains a stationary device at one end which causes the fluid to swirl and rotate about the centerline of the meter. At the opposite end of the meter body, another device restores the original fluid motion. Because the geometry of the flow area of the meter varies, the center of the fluid rotation moves (precesses) and the frequency of precession is related to the gas flow rate. An electronic circuit including a thermistor and voltage measurement is used to convert the precession to a flow rate.

In many cases when an obstruction is placed in a flowing fluid, vortices (plural of vortex) are shed behind or at the downstream part of the obstruction (Figure 9-20). Evidence of the vortices can usually be seen, for instance, behind an oar or paddle moved through water. The rate of formation and shedding of these vortices can be related to the overall fluid flow rate. Differences in pressure occur as the vortices move, which

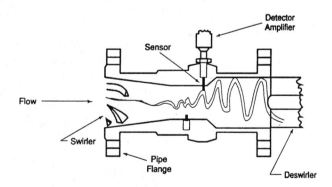

Figure 9-19 Vortex flow meter cross-section *(From Jack W. Chaplin,* Instrumentation and Automation for Manufacturing, *© 1992, Delmar Publishers, Inc., Albany, N.Y.)*

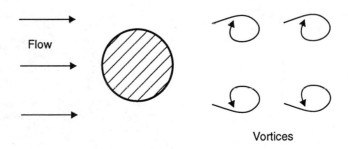

Figure 9-20 Vortices form behind an object as the fluid passes.

can be used to actuate a sensor at a rate proportional to the flow rate. In an *ultrasonic* meter, a cylinder is set perpendicular to the fluid motion. An ultrasonic transmitter is positioned downstream from the cylinder. The transmission of the ultrasonic signal is affected by the number of vortices leaving the cylinder and can be correlated to the fluid flow rate.

Measurements relating to heat can be used to measure flow rates. In one type of application, the amount of heat required to raise the temperature of laminarly flowing fluid by a certain amount is measured. In another type, either the temperature of a device is held constant and the amount of heat required measured, or the temperature of the device is measured as the amount of input heat is held constant. The latter application is used primarily with gases. This method is sometimes referred to as *hot-wire anemometry*. An advantage of these methods is that they do not appreciably obstruct flow and have no moving parts.

Another type of flow meter makes use of the force acting on an object (Equation 9-20). The object, usually a disk, is placed in the flow and the resulting force is transmitted to a supporting device which, in turn, is transmitted to a force transducer, such as a strain gage arrangement. This type of device can be used on liquids or gases and can also handle condensates.

Open-Channel Flow. If the fluid flow is not in a closed conduit, it is called *open-channel* flow. Such flow is said to have a *free surface*, that is, a surface in contact with the ambient atmosphere. One of the most common devices for measuring this type of flow is a *weir* (Figure 9-21). Weirs have openings through which the fluid passes. The height of the fluid in the opening can be related to the flow rate. *Flumes* offer an alternative to weirs in that they require less head (energy) loss for operation. One of the more popular flumes, called Parshall,

Figure 9-21 Weir for measuring open-channel flow *(From Jack W. Chaplin,* Instrumentation and Automation for Manufacturing, *© 1992, Delmar Publishers, Inc., Albany, N.Y.)*

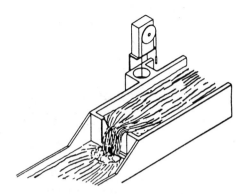

resembles to some extent an open-topped square cross-section Venturi tube (Figure 9-22). Again, the height of the fluid in the flume can be related to the flow rate. An *open nozzle* can also be used to measure flow rate (Figure 9-23). A common application of one of the three open-channel devices includes a still well adjoining and connected to the device. The fluid level in the still well can be correlated with the flow rate.

Figure 9-22 Parshall flume for measuring open-channel flow *(From Jack W. Chaplin,* Instrumentation and Automation for Manufacturing, *© 1992, Delmar Publishers, Inc., Albany, N.Y.)*

Figure 9-23 Open flow nozzle for measuring open-channel flow *(From Jack W. Chaplin,* Instrumentation and Automation for Manufacturing, *© 1992, Delmar Publishers, Inc., Albany, N.Y.)*

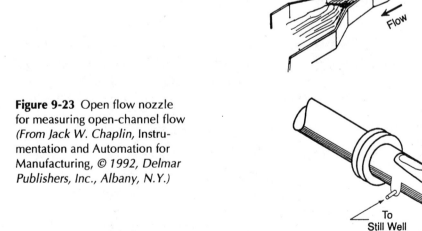

Total Flow

A separate class of devices is used to measure total flow. These devices are also referred to as *quantity* meters.

Positive-Displacement Meters. Positive-displacement meters operate by filling a chamber of known volume with fluid and then emptying the chamber. The number of times the chamber is filled and emptied in a given length of time is a measure of total flow. These devices can have more than one chamber; thus, the total number of chamber fillings and emptyings must be counted. Note that the average flow rate can be calculated by dividing the total flow by the given time interval. In a *piston* meter, fluid enters on one side of the piston; and when the volume is full, valves open and the piston reverses direction. Fluid now enters the meter on the opposite side of the piston. In this way, the volumes on the two sides of the piston fill and empty; and the number of these happenings can be used to determine total flow. In a *nutating* disk meter (sometimes referred to as a wobble meter), the fluid moves through the chambers and causes a disk to rotate and wobble (Figure 9-24). This motion can be transferred to a counting device. Other types of positive-displacement meters include rotary vanes and gears.

Velocity Meters. Devices used for measuring flow rate can be outfitted such that they measure total flow. The velocity-sensing portion of the flow rate meter can be connected to a mechanism that totalizes flow.

Mass Flow

If the density (or alternatively, the specific weight) of the fluid is also measured, the mass flow can be measured. Designs for mass flow measurement include constant speed impeller-

Figure 9-24 Nutating disk flow meter cross-section *(From Jack W. Chaplin, Instrumentation and Automation for Manufacturing, © 1992, Delmar Publishers, Inc., Albany, N.Y.)*

turbine wheel–spring combinations that relate the spring force to mass flow and devices that relate the movement of heat (change in temperature) to mass flow.

APPLICATION CONSIDERATIONS

The reader will note that flow rate and total flow measurement involve many different devices and principles of operation. It is not surprising that there are also a number of items to be considered when making applications.

Selection

Considering the pressure differential devices, the cost of purchase and installation is least for the orifice plates, followed by the flow nozzle and then the Venturi tube. The amount of energy dissipated and the accuracy that can be expected are in a reverse order; that is, the Venturi tube is the most accurate and least energy destroying. The Venturi tube and flow nozzle are best if the fluid carries an appreciable amount of sediment. Differential pressure meters are generally recommended for use in the 30% to full-flow range. At less than 30% full flow, measurement errors may become too large.

Turbine meters are usually limited to service with clean fluids with low viscosity, such as alcohol or gasoline. They can be used when vapor is present with the fluid.

The accuracy of most magnetic flow meter installations is within ±1% of the full-scale reading. These meters are excellent for measuring flow rates of fluids with high viscosities.

For open-channel flow, the flume is preferable to the weir if the fluid contains large amounts of suspended materials. In general, the flume is more costly to obtain and install.

Installation

For orifice plates using pipe taps, the upstream tap should be 2.5 pipe diameters from the plate, and the downstream tap approximately 8 diameters away. With vena contracta taps, the upstream tap should be one pipe diameter from the plate and the downstream tap located according to the manufacturer's recommendations for a particular installation.

Taps for a Venturi tube should be at locations corresponding to the minimum and maximum conduit diameters. For a flow nozzle, the manufacturer's recommendation for tap locations should be carefully followed.

For most devices, it is desirable to have a straight run of pipe upstream from the device to allow pressure fluctuations from flow direction changes to even out. This is particularly true for orifice plate application where straight-pipe runs of at least ten times the pipe diameter are recommended both upstream and downstream from the plate. It is sometimes advantageous to use flow-straightening vanes to align flow.

A flow nozzle should be installed in a straight section of pipe, and it should be kept as far downstream as possible from any major flow disturbance. The nozzle should be installed vertically upward if there are vapors or gases in the fluid and vertically downward if there is suspended material.

Special attention is called to use of a Pitot-static tube in that there should be at least ten pipe diameters of straight pipe run upstream and downstream from the point of measurement. Even then, flow straighteners are recommended.

In general, flow rate should not be measured while the flow is pulsing. Pressure differences resulting from acceleration or deceleration of the fluid make interpretation of the pressure difference readings difficult. It is, however, possible to design and install damping devices to aid in coping with pulsations.

Calibration

Flow meter calibration is usually carried out using a constant flow rate and some method for determining total flow over a specified period of time. It is also fairly common to calibrate a meter by putting it in series with a meter that has been previously calibrated. The latter procedure is called *direct secondary* calibration.

Some meter readings can be made more accurate in certain portions of their range by making a correction based on Reynolds numbers. Literature is available for applying such corrections.

Finally, methods are available for making corrections necessary if the flowing fluid temperature is different than that for which the meter calibration was made. These methods can also be used to convert flow rate readings to equivalent flow rate at a reference temperature as is required in some industries.

REVIEW MATERIALS

Important Terms

velocity
static pressure
dynamic pressure
laminar flow
parabolic velocity
 distribution
turbulent flow
boundary layer
Reynolds number
viscosity
Bernoulli equation
flow rate meters
total flow meters
continuity equation
head loss
friction factor
minor losses
form drag
open nozzle
quantity meter
piston meter
nutating disk meter
velocity meter

concentric plate
eccentric plate
segmental plate
flange tap
pipe tap
vena contracta tap
beta ratio
Pitot-static tube
rotameter
turbine flow meter
electromagnetic flow meter
vortex precession meter
ultrasonic meter
hot-wire anemometry
open-channel flow
free surface
flume
drag coefficient
mass flow meters
orifice plate
Venturi tube
flow nozzle

Questions

1. How does a positive-displacement meter work?
2. Discuss the advantages and disadvantages of orifice plates, Venturi tubes, and flow nozzles for use in measuring flow rate.
3. What might be a reason for not selecting an elbow flow-measuring device?
4. What is the primary use of the Reynolds number for describing fluid flow?
5. What is the difference between total flow meters and mass flow meters?
6. Why is it necessary to have straight sections of pipe upstream from some flow meters?
7. What effect might be observed if pressure tap fittings project into the flowing fluid?
8. What are the two basic types of flow measurement, and what is the difference between them?
9. Sometimes a rotameter is called a variable-area flow meter. Why?

10. Discuss three devices for measuring open-channel flow.
11. How can a total flow meter be used to estimate flow rate?
12. What properties of a fluid affect the way it flows in a pipe?
13. What is the difference between laminar and turbulent flow?
14. Why is there less head (energy) loss with a Venturi tube than with an orifice plate?
15. What is the beta ratio, and how does it affect flow rate measurement?
16. How is a flow nozzle similar to and different from a Venturi tube?
17. Why are orifice plates one of the most popular methods for measuring flow rate?
18. When would a segmented orifice plate be used in place of a concentric one?
19. How can a Pitot-static tube be used to measure flow rate?
20. Describe the operation of an electromagnetic flow meter.

Problems

1. Five hundred gallons per minute flow through a pipe with an internal diameter of 12 inches. Downstream the inside diameter is 6 inches. What is the velocity of flow in each section?
2. In Problem 1, what is the equivalent m^3/sec?
3. Five hundred liters per minute flow through a pipe with an internal diameter of 850 mm. What is the weight flow rate for water with a specific weight of 9,810 N/m^3?
4. If one section of a pipe has a diameter four times larger than the diameter at another section, what is the ratio of the average velocities at the two sections if there is steady flow?
5. Ten gallons per minute flow through a pipe with an inside diameter of 2 inches. What is the average velocity in the pipe?
6. What is the weight flow for Problem 5 if water is flowing?
7. A certain process requires 3,000 liters of water per minute. How many m^3/sec is this?
8. Four gallons of water per second flow through a pipe with an inside diameter of 3 inches. What is the average velocity and weight flow rate?
9. Water flows at 3 cfs in a pipe with an inside diameter of 3 inches. The pipe diameter decreases to 2 inches. What is the average velocity in each section?
10. A pipe with an inside diameter of 6 inches carries water at the rate of 160 cfm. It branches into two lines, one of which carries 100 cfm. If the other branch has an inside diameter of 6 inches, what is the average velocity inside it?
11. A large open-top tank has an opening 4 feet below the surface of the water in the tank. If the area of the opening is 4 in^2, what is the velocity of the water leaving the hole?
12. Water leaves a nozzle going straight up vertically at a velocity of 50 fps. What is the maximum height it could reach if there were no air resistance?

13. Water falls over a dam 300 feet high. What would its velocity be just before striking the bottom if air resistance is ignored?
14. A horizontal pipe decreases from an inside diameter of 4 inches to 2 inches. If the pressure in the 4 inch section is 100 psig and 400 gallons per minute of water are flowing, what is the pressure in the 2 inch section?
15. If the pipe orientation in Problem 14 is changed to vertical with flow upward from the 4 inch to the 2 inch section, what is the pressure at a location in the 2 inch section 20 feet above a point in the 4 inch section? Neglect all losses.
16. Water flows in the pipe at a rate of 60 cfm. What is the magnitude of h if the pressure at A is 50 psig? (See Figure 9-25.)

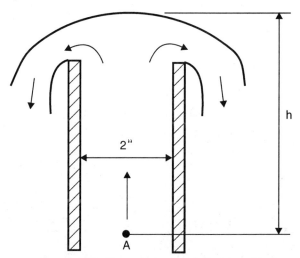

Figure 9-25 Figure for use with Problem 16.

17. Oil with a specific gravity of 0.8 flows in a horizontal pipe at the rate of 2,000 gallons per minute. Pressure at an upstream section with an inside diameter of 7 inches is 50 psig. If the inside diameter of the pipe is 3.5 inches at a downstream section, what is the pressure there? Neglect all losses.
18. Seawater with a specific weight of 64.0 lbs/ft^3 is at a level of 4 feet above the centerline of the exit pipe. The tank is closed at the top, and there is a pressure of 1 psig above the seawater. What is the velocity at the end of the pipe? (See Figure 9-26.)
19. A Pitot-static tube is used to measure the velocity of water. If a pressure difference of 2.5 inches of water is measured with a manometer, what is the velocity?
20. Water flows through a horizontal pipe with a cross-sectional area of 1.44 in^2. At a section upstream, the area is 2.88 in^2. The difference in pressure between a point in the larger section and a point in the smaller section is 0.096 psi? What is the flow rate in the pipe? Neglect all losses.

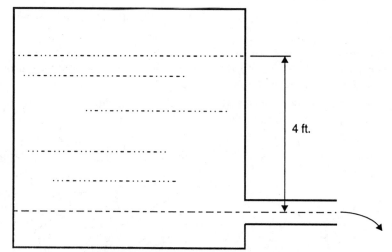

Figure 9-26 Figure for use with Problem 18.

21. What is the head loss for 40 feet of straight, horizontal, laminar flow in a pipe with an inside diameter of 1 inch? The fluid specific weight is 50 lbs/ft³, the viscosity is 1.05 x 10⁻⁵ lb sec/ft², and the average viscosity is 2 fps.

22. What is the head loss for 500 feet of pipe with an inside diameter of 1 inch if the average fluid velocity is 3 fps and the friction factor is 0.03?

23. If the K values for the valves, elbows, and other fittings in a 1 inch diameter flow line system add up to 24, what length of pipe would give an equivalent head loss if the friction factor is 0.025?

24. What is the drag force on a sphere with a 4 inch diameter if it moves through water at a rate of 10 fps? Assume a drag coefficient of 0.3.

25. A circular disk faces moving fluid with a specific weight 64.4 lbs/ft³. The area of the disk is 0.3 ft². If the drag coefficient is 1.0 and the force on the disk is 30 lbs, what is the fluid velocity?

26. A positive-displacement meter has four chambers, each having 25 in³ volume. If a counter indicates that the meter chamber assembly rotates 400 times in 5 minutes, what are the total flow and average flow rate over the 5 minutes?

27. A total flow device indicates that 1,400 gallons have passed over three separate one-hour periods. What was the average flow rate during the time of flow?

28. What would be an acceptable range of orifice plate hole sizes if the inside diameter of the pipe for which they were to be applied was 8 inches?

29. The flow of a fluid in a pipe with an inside diameter of ½ inch must be laminar. The fluid kinematic viscosity is 2 x 10⁻⁵ ft²/sec. What is the largest average fluid velocity that can be used? Use Reynolds number of the following form:

$$R = \frac{VD}{v}$$

30. Fluid with a kinematic viscosity of 1×10^{-5} ft²/sec flows in a tube with an inside diameter of ⅛ inch at 10 fps. Is the flow laminar or turbulent? Use Reynolds number of the following form:

$$R = \frac{VD}{v}$$

10

Humidity

CHAPTER GOALS

After completing study of this chapter, you should be able to do the following:

Understand what humidity is and the different ways of expressing it.

Calculate the relative humidity and the specific humidity or humidity ratio for a given set of conditions.

Describe the difference between dry-bulb and wet-bulb temperatures and what the term dew point means.

Use a psychrometric chart to estimate such factors as relative humidity, amount of moisture in the air, dry-bulb temperature, wet-bulb temperature, dew point, heat content, and specific volume.

Understand the operation of devices for measuring humidity, dew point, and moisture content.

Know other considerations involved with applying humidity-measuring devices.

The amount of moisture in the air or the humidity is important in industrial operations for its effects on such things as automatic controls, ingredient mixing, and the comfort of humans involved. Specific industries prone to its effects include pharmaceuticals, textile mills, and printing operations.

BASIC CONSIDERATIONS

Basic Terms

The term *humidity* refers to the amount of water vapor present in the air or in other gases. By itself, humidity is a rather general term; and to describe the particular item of interest, other words are used in conjunction. The following sections describe several "humidities" and terms used in their description and measurement.

Relative Humidity. *Relative humidity*, often represented by ϕ, is the amount or quantity of water vapor present in a given volume expressed as a percentage of the amount or quantity of water vapor that would be present in the same volume under saturated conditions at the same temperature. *Saturated conditions* means that it would not be possible for more water vapor to be present without condensation taking place or tiny water droplets forming and collecting on the surfaces of the container. In formula form,

$$\text{Relative humidity} = \frac{\text{amount of water vapor present in given volume of air or gas}}{\text{maximum amount of water vapor that the volume of air or gas could hold at that temperature and pressure}} \times 100 \quad (10\text{-}1)$$

Another definition for relative humidity is given in terms of pressures:

$$\text{Relative humidity} = \frac{\text{water vapor pressure in air or gas}}{\text{water vapor pressure in saturated air or gas at that temperature}} \times 100 \quad (10\text{-}2)$$

As the above formulas indicate, a relative humidity of 100% means that the air or gas cannot "hold" any more water vapor. An example of 100% humidity is the way one can "see" one's breath on a cool morning. What is really seen are tiny droplets of water forming from the saturated water vapor in the air.

Specific Humidity and Humidity Ratio. The terms *specific humidity* and *humidity ratio* are synonymous. Having been developed in separate areas of application, they both are defined as follows:

$$\text{Humidity ratio} = \frac{\text{mass of water vapor in mixture}}{\text{mass of dry air or gas in mixture}} \quad (10\text{-}3)$$

The units for this ratio are lbs water vapor per lb dry air or gas. Occasionally, this ratio may be referred to as absolute humidity. Although the development of the following formula is not important as far as this chapter is concerned, the formula itself may aid in the understanding of the humidity ratio and in making calculations.

$$\text{Humidity ratio} = \frac{\text{mass}_{\text{water vapor}}}{\text{mass}_{\text{air or gas}}} = \frac{0.622\, p_{\text{water vapor}}}{p_{\text{mixture}} - p_{\text{water vapor}}} \quad (10\text{-}4)$$

In this formula, the p's represent pressure and "mixture" refers to the water vapor-air or gas combination. It can further be shown that

$$\text{Humidity ratio} = 0.622 \left(\frac{p_{\text{water vapor}}}{p_{\text{air or gas}}} \right) \quad (10\text{-}5)$$

noting that p_{air} or p_{gas} is a partial pressure.

Dry-Bulb Temperature. Some of the calculations relating to humidity involve the *dry-bulb temperature*. By definition, this is the temperature of the water vapor–air or gas mixture as measured by a thermometer whose sensing element is dry. Thus, a thermometer mounted on the wall of a room measures a dry bulb temperature.

Wet-Bulb Temperature. The *wet-bulb temperature* is measured at essentially the same location as the dry-bulb temperature; but, in contrast to the dry-bulb temperature, it is taken with a thermometer whose sensing element is kept moist. Usually, provision is made to cause the air or gas to move past the wet bulb so that vaporization takes place. Because heat is required for vaporization to take place, the sensing element is cooled; thus, its reading is called the wet-bulb temperature.

Dew Point. The *dew point* is the temperature at which condensation (dew) will begin to form from the moist mixture as the mixture is cooled at a constant pressure. A more technical definition of dew point is the saturation temperature of the mixture at the corresponding vapor pressure.

Psychrometric Chart. Many calculations and estimations relating to humidity can be made easier through the use of a *psychrometric chart*. As shown in Figure 10-1, this is, at first glance, a rather complicated device. An inspection of the variables, one by one, will show that its use need not be overwhelming. The horizontal axis represents the dry-bulb temperature of the mixture in degrees Fahrenheit. The next scale to look at is the uppermost curved one leading upward and to the right. It is labeled "Wet-Bulb and Dew Point Temperatures." Note that the numbers on this scale are associated with straight lines leading downward to the right. Before considering an example, note the rest of the lines that curve upward and to the right as does the "Wet-Bulb and Dew Point Temperature" scale. On these curves, note the different percentage relative humidity values.

As a first example, assume that a dry-bulb temperature of 80° F was measured along with a wet bulb temperature of 60° F. Following the 60° wet-bulb line downward and to the right until it intersects the vertical line representing the dry-bulb temperature of 80°, it can be seen that the intersection takes place on the 30% relative humidity line. Thus, a room temperature of 80° F and a wet-bulb temperature of 60° F means that the relative humidity in the room is 30%.

Further study of the relative humidity curves shows that as one moves upward, the percentage relative humidity increases until the last curve (the one with the "Wet-Bulb and Dew Point Temperatures") is reached. Notice that this outermost curve is the 100% relative humidity line; that is, the air can hold no more water vapor. This also means that condensation or dew would begin to form and, therefore, the term dew point temperatures. To further illustrate, what would the wet-bulb temperature be if the dry-bulb temperature were 90° F and the

Figure 10-1 Psychrometric chart for air–water vapor mixtures *(Courtesy of General Electric)*

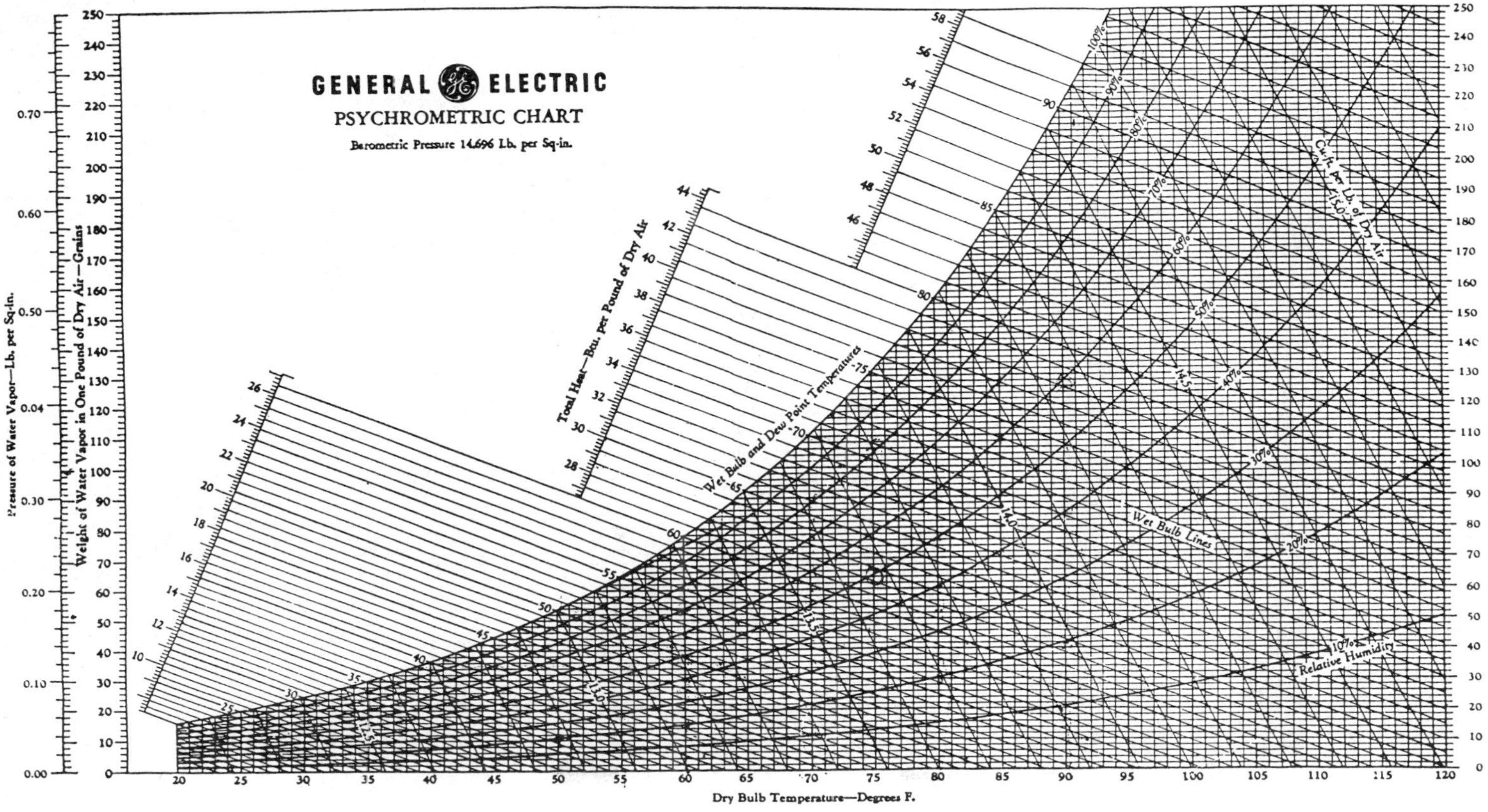

relative humidity were 100%? Checking the chart, one sees that the wet-bulb temperature is 90° F. Indeed, that is true; at 100% relative humidity, the dry-bulb and wet-bulb temperatures are the same. The following examples illustrate the use of other information from the psychrometric chart.

EXAMPLE 10-1

Assume that near the end of a day, the outdoor temperature is 60° F and the relative humidity is 50%. How much would the temperature have to drop for dew to form assuming no other weather conditions change?

Note the vertical scale on the left side of the psychrometric chart labeled "Weight of Water Vapor in One Pound of Dry Air-Grains." Under the conditions of the question, the amount of moisture in the air will remain the same as the temperature drops. The first step in the solution is to find the intersection of the 60° F dry-bulb temperature line with the 50% relative humidity line. Because the amount of moisture remains constant, move horizontally to the left until reaching the "Dew Point Temperature." The value is approximately 42° F—the temperature at which dew would begin to form.

EXAMPLE 10-2

How much more moisture does air at 90° F and 40% relative humidity hold as compared to 90° F and 20% relative humidity?

From the chart at 90° F and 40% relative humidity, moving horizontally to the left, one reads about 84 grains of water vapor per pound of dry air. (There are 7,000 grains in one pound.) At 90° F and 20% relative humidity, the corresponding number is about 42 grains per pound. Thus, in this case, doubling the relative humidity (at constant temperature) doubles the amount of water vapor in the air.

The psychrometric chart shown in Figure 10-1 is for the standard atmospheric pressure of 14.696 psi. For other atmospheric pressures, the numbers would be slightly different. That part of the atmospheric pressure resulting from the water vapor is given on the outermost left vertical axis. At 80° F and 35% relative humidity, for example, the water vapor pressure is about 0.17 psi. Since the atmospheric pressure is made up of the water vapor pressure and the "air" pressure, the part of the atmospheric pressure due to the air is 14.696 - 0.17, or approximately 14.53 psi.

Another item to be noted on the psychrometric chart is the "Total Heat—BTUs per Pound of Dry Air." Information of this type is useful in calculating the amount of heat needed to warm a given amount of air by a certain number of degrees (or the amount of heat available when cooling a given amount of air by a certain number of degrees). Note that the lines used are the same as those for wet-bulb and dew point.

EXAMPLE 10-3

How much heat must be supplied to heat air at 40° F and 70% relative humidity to 80° F and 50% relative humidity?

From the chart at 40° F and 70% relative humidity, move upwardly and to the left parallel to the total heat lines and read about 12.5 BTUs per pound. The corresponding number for 80° F and 50% relative humidity is about 31 BTUs per pound. The difference, or 31 − 12.5 = 18.5 BTUs, would be the amount of heat to be added.

The final lines to note from the psychrometric chart are those running almost vertically and labeled "Cu ft per lb of Dry Air." As the units indicate, the lines give the "space" required for one pound of dry air. Thus, at 70° F and 40% relative humidity, just less than 13.5 ft^3 contain one pound of dry air.

MEASURING DEVICES

Hygrometers

Hygrometers measure relative humidity making use of the fact that physical or electrical changes take place in certain materials as they absorb moisture. Examples of these materials include hair, skin or membrane, and thin strips of wood attached to a metal strip and wound in the shape of a helix. As these materials "pick up" moisture, they change their length or geometrical properties. The amount of change can be related to the change in relative humidity. In general, these types of hygrometers can be used between 20% and 90% relative humidity and for temperatures up to 150° to 160° F. Their error can be rather large, in the order of 5% relative humidity.

One type of electrical device uses a capacitance probe. A strip of aluminum with an oxide layer is coated with a thin film of gold. The aluminum and gold serve as capacitor plates, and the oxide layer serves as a dielectric. As water vapor condenses and collects on the oxide layer, the capacitance changes.

Another type of electrical device uses a change in resistance to indicate relative humidity. The resistance element is made up of a double winding of wires on a form made from an insulating material. These two windings have a definite gap between them. Lithium chloride is used to coat the wires and sets up a conducting path between them. As the amount of moisture in the air varies, the resistance of the unit varies and can be related to the relative humidity. This type of device is readily suited to continuous recording or for control purposes.

Because certain salts are *hygroscopic*—that is, their water content varies with humidity—devices such as those shown in Figures 10-2 and 10-3 can be used. The changing of the water content causes a change in the salt's conductivity.

Psychrometer

Psychrometers are devices for determining relative humidity using temperature measurements and a chart. In essence, it is the vaporization rate that is used to furnish information for determining the relative humidity. Two temperatures are

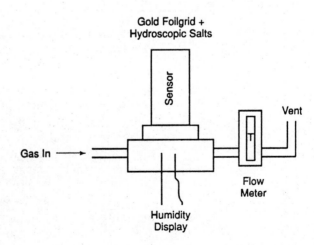

Figure 10-2 Hygrometer operation based on conductivity change in hydroscopic salts *(From Jack W. Chaplin,* Instrumentation and Automation for Manufacturing, *© 1992, Delmar Publishers, Inc., Albany, N.Y.)*

Figure 10-3 Electrolytic hygrometer sensor *(Courtesy of General Eastern Instruments Co.)*

measured: the dry-bulb and wet-bulb. The dry-bulb thermometer is an ordinary temperature-sensing device and is so called to differentiate it from the wet-bulb thermometer. The sensing element of the wet-bulb thermometer is kept wet so as to set up the decrease in temperature associated with vaporization. If the air surrounding the wet-bulb sensing device is of rather low relative humidity, the vaporization rate from the wet bulb is rather high. If the air has a high relative humidity, the vaporization rate will be low. Since vaporization requires heat, the loss of heat from the wet-bulb sensing element causes it to become cooler. In this way, the rate of vaporization (and hence the relative humidity) can be related to the wet-bulb temperature.

One of the most important requirements is that the air surrounding the wet-bulb element be changed at such a rate that the air near the element does not become saturated with water vapor. One of the devices developed to provide the necessary air movement is called a *sling psychrometer*. As shown in Figure 10-4, two thermometers are mounted on a frame which can be rotated about a handle. One of the thermometer bulbs is covered with a soft porous material completely saturated with water. When the device is rotated, the wet bulb will be cooled by the water vaporization and the dry-bulb reading will remain constant. Usually, the device is rotated for 15 to 25 seconds and then the wet-bulb temperature reading made quickly before it begins to rise again. The dry-bulb and wet-bulb readings can be taken to the psychrometric chart and the appropriate numbers obtained.

The sling psychrometer principle can be incorporated into a recording device. The temperature sensors used are usually pressure-spring or resistance bulbs.

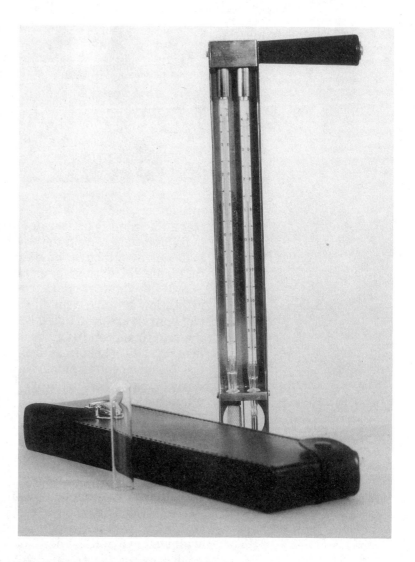

Figure 10-4 Sling psychrometer *(Courtesy of Qualimetrics, Inc., A Dynatech Company)*

Dew Point–Measuring Devices

One of the simplest devices for measuring dew point consists of a smooth, clean cylinder through which cold water is circulated. The temperature of the cylinder when a film of moisture begins to form on its outer surface gives the dew point of the air. Methods are available for monitoring and controlling the temperature of the circulating water and the detection of the moisture film. Similar systems are available which replace the circulating water with gases that are expanding and thus cool the cylinder.

Another method for determining dew point consists of a mirror surface bonded to a thermistor holder. The gas whose dew point is being measured is passed by the mirror. Light from a lamp beamed at the mirror is reflected toward a photoelectric resistor. When dew forms on the mirror, there is a change in the amount of light reflected. This change is detected by an optical sensing bridge. The thermistor thus senses the temperature at which dew forms. Figure 10-5 illustrates this method.

Another dew point device consists of a nonconducting cylinder with two evenly spaced windings of wire. The wires rest on a fabric impregnated with lithium chloride which has previously been placed on the cylinder. The general operation of this device is similar to that noted previously. This particular device can be assembled in such a way as to provide continuous recording of the dew point.

Moisture Content

One of the methods for determining moisture content is based on the fact that water absorbs microwaves. As the microwaves are absorbed, their amplitude and phase are changed. These

Figure 10-5 Dew point sensor *(Courtesy of General Eastern Instruments Co.)*

changes can be related to the moisture content of the gas containing the moisture. A typical measuring instrument consists of a transmitter, receiver, and two directivity horns. Because of the microwaves' ability to travel through various materials, the instruments can be mounted in a variety of ways, depending on how the material containing the moisture is moving. Application is possible for tanks, chutes, belts, pipes, and the like. Operation is straightforward; microwaves leave the transmitter, pass through the material of interest, and are then sensed by the receiver and recorded.

A second method of measuring moisture content is with infrared waves. Their use is based on the *reflectance* of the material under consideration, that is, its ability to absorb and scatter the infrared waves. The reflectance of any given material is dependent on its chemical composition and moisture content. Operation of a device using this method consists of projecting the waves onto the surface of the material indirectly through a system of lenses and mirrors. The purpose of the lenses is to distribute uniformly the waves on the mirrors. Energy reflected by the material is deflected by another mirror to a cell which converts the radiant energy into electrical energy. Measured wavelengths and amplitudes and reference wavelengths and amplitudes are used to produce two currents. The difference between these currents can be related to moisture content.

Other types of moisture-measurement methods include color change, titration, neutron reflection, and nuclear magnetic resonance.

APPLICATION CONSIDERATIONS

The use of the sling psychrometer should follow as closely as possible to the manufacturer's recommendation. Distilled water should be used for the wet bulb, and it should be as close as possible to the ambient temperature. Movement of the wet bulb and its orientation should be such that air near the wet bulb does not reach a "too saturated" condition. The dry bulb should be kept free from water or other liquids. Any measurements taken outdoors should be made in the shade.

One of the best standards for calibration of humidity measuring devices is the National Bureau of Standards gravimetric hygrometer. In this laboratory device, water vapor in an air sample is absorbed by chemicals and then weighed.

REVIEW MATERIALS

Important Terms

humidity
relative humidity
saturated conditions
specific humidity
humidity ratio
dry-bulb temperature
wet-bulb temperature

dew point
psychrometric chart
hygrometer
hygroscopic
psychrometer
sling psychrometer
reflectance

Questions

1. Which "humidity" is the one given during weathercasts, and what is its significance for daily activity?
2. If the dry-bulb and wet-bulb temperatures are known, what information can be obtained from the psychrometric chart?
3. What is the difference between a psychrometer and a hygrometer?
4. How does a sling psychrometer work?
5. What is the function of the lithium chloride in a moisture-sensing device?
6. What property of water allows for measurement of moisture with microwaves?
7. How would a "hair" element function in a humidity measuring device?
8. When a wet-bulb/dry-bulb device is used to measure humidity, which bulb will have the higher temperature? Why?
9. How might humidity measurements be incorrect even though the instruments are in good condition and being correctly used?
10. Why should the water supply for the wet bulb be at ambient temperature?

Problems

1. For a dry-bulb temperature of 70° F, what is the relative humidity when the difference in the wet- and dry-bulb readings is 5°? 10°? 20°?
2. If the wet-bulb temperature is 65° F, what is the relative humidity when the difference in the wet- and dry-bulb readings is 5°? 10°? 20°?
3. What would be the dew point for a condition of 80° F (dry bulb) and 50% relative humidity? 80° F (dry bulb) and 80% relative humidity?
4. If the dew point is 70° F, what temperature corresponds to a relative humidity of 30%?
5. What are the absolute humidities corresponding to Problems 3 and 4?

6. What relative and absolute humidities correspond to a dry-bulb temperature of 75° F and a wet-bulb temperature of 65° F?
7. What relative and absolute humidities correspond to a dew point temperature of 83° F and a dry-bulb temperature of 106° F?
8. What relative humidities correspond to the following temperatures and dew points?

Temperature, °F	Dew Point, °F
30	25
70	58.5
110	91.7

9. What is the pressure of the water vapor in the air when the temperature is 70° F and the relative humidity is 40%? 70° F and 80%?
10. How much heat is required to raise the temperature of one pound of dry air from 40° F to 70° F if the relative humidity is constant at 50%?
11. How much heat does it take to heat one pound of dry air if the temperature goes from 40° F to 80° F and the relative humidity is constant at 60%?
12. How much heat is required to raise the temperature of one pound of dry air if the temperature of the air is constant at 70° F and the relative humidity goes from 25% to 75%?
13. How much water would have to be added to one pound of dry air at 80° F to increase the relative humidity from 20% to 70%?
14. If air is "dried" from 90% relative humidity to 30% relative humidity with a constant temperature of 60° F, how much water is removed?
15. Air is "conditioned" from 95% relative humidity and 90° F to 40% relative humidity and 74° F. How much heat and water are removed per pound of dry air?

Other Variables

CHAPTER GOALS

After completing study of this chapter, you should be able to do the following:

Describe the meaning of density and specific gravity.

Understand how the calculation of the specific gravity of a gas differs from the calculation of the specific gravity of a solid or liquid.

Understand the operation of devices used for measuring density or specific gravity and some of the items involved with their applications.

Know the difference in the different ways of expressing viscosity.

Describe different devices for measuring viscosity and their applications.

Understand the terminology involved with measuring position and displacement and some of the measuring devices available.

Describe the difference between force and torque, make calculations, and describe measuring devices available.

Understand terms involved with the sound phenomenon, such as noise, random noise, primary and secondary sources, standing waves, decibel, and loudness.

Describe measuring devices associated with sound and their applications.

Understand the concept of solution acidity or alkalinity in terms of pH values.

Describe pH-measuring devices and their applications.

There are several variables other than those covered in the preceding chapters which can be of importance in an industrial or manufacturing situation. Because the importance of these variables is sometimes less than those already covered—and, in some cases, less space is necessary to cover them—the format for this chapter is different from that of the preceding ones. In this chapter, the main headings indicate the variable to be discussed. The subheadings will cover the basic terms, measuring devices, and application considerations.

DENSITY AND SPECIFIC GRAVITY

Basic Terms

The *density* of a material is defined as its mass per unit volume. It is usually represented by the Greek letter ρ. There is a relationship between density, specific weight, and the acceleration of gravity of the following form:

$$\gamma = \rho g \qquad (11\text{-}1)$$

where γ is the specific weight of the material in lbs/ft^3 and g is the acceleration of gravity in ft/sec^2. The units of density in this form are slugs/ft^3.

The specific gravity of a liquid is the ratio of its specific weight (or density) and the specific weight (or density) of water at a standard temperature. Thus, because mercury has a specific weight of approximately 848.6 lbs/ft^3, its specific gravity is $848.6/62.4 = 13.6$. Note that specific gravity has no units. Another way to perceive specific gravity is that it compares the weight of a given volume of a material with the same volume of water. The specific gravity of a gas is the ratio of its specific weight (or density) to that of air at 60° F and 14.7 psia (standard conditions).

Measuring Devices

A *hydrometer* is probably the most common device for measuring liquid density. As shown in Figure 11-1, the hydrometer floats in the liquid whose density is being measured and, because of buoyancy, displaces a volume of liquid equal to its own weight. The hydrometer is usually made of hollow glass

Figure 11-1 The level at which the hydrometer floats is an indication of the liquid's specific gravity or density.

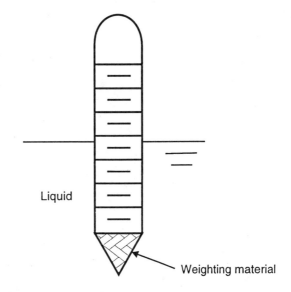

Figure 11-2 Thermohydrometer.

or metal and contains a weighting material at one end that causes it to float in an upright position. The greater the density of the liquid, the higher the equilibrium position of the float. The liquid density is read from a scale on the outside of the hydrometer. Because the density of most liquids varies with temperature and not all liquids expand at the same rate, it may be necessary for those cases where a high degree of accuracy is desired to monitor the liquid temperature and make corrections from a standard reference. Figure 11-2 shows a thermohydrometer that contains a thermometer to assist in making corrections.

Remote reading of a hydrometer can be made possible by using a metal rod as a weight and then using the rod along with a coil as an induction circuit. The entire system must be carefully designed and assembled, but it is possible to set up a continuously measuring hydrometer in this way (see Figure 11-3).

Liquid density can be measured by weighing a fixed volume of the liquid. Continuous measurement can be made if care is taken to assure that liquid flow rates into and out of the volume are the same.

Liquid density can also be measured with a displacement device. The displacement element is enclosed in a fixed-volume chamber; and as the density of the liquid changes, the buoyant force on the element changes. The element is also

Figure 11-3 Induction hydrometer *(From Jack W. Chaplin, Instrumentation and Automation for Manufacturing, © 1992, Delmar Publishers, Inc., Albany, N.Y.)*

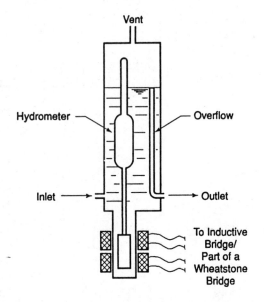

attached to a mechanism which then converts the force into a density reading.

The change in the vibration of an object in contact with the fluid under consideration is related to the density of the fluid (see Figure 11-4). Power to drive the vibrating element is from an outside source. As the fluid density increases, the vibration decreases. In a nonpowered application, the element can be tapped and the resulting frequency observed.

Pressure or difference in pressure measurement as described in Chapter 5 can be used to measure density. A single pressure measurement can be used in a constant-level ar-

Figure 11-4 Specific gravity and density can be measured with a vibration-sensing device *(From Jack W. Chaplin, Instrumentation and Automation for Manufacturing, © 1992, Delmar Publishers, Inc., Albany, N.Y.)*

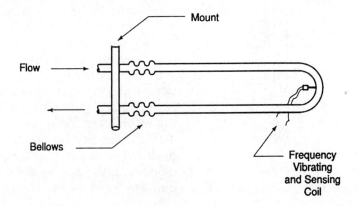

Figure 11-5 Liquid density can be obtained by measuring the pressure at the base of a constant-level tank.

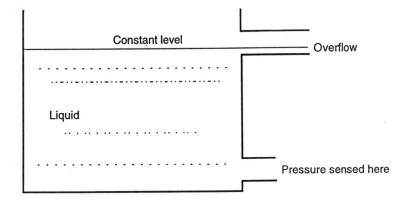

rangement as shown in Figure 11-5. For either an open or closed tank with varying liquid level or pressure (or both), two pressure measurements are necessary to obtain the pressure difference. The basic formula to be used with this method is as follows:

$$\rho = \frac{p \text{ or } \Delta p}{gh} \quad (11\text{-}2)$$

where ρ has the units noted earlier and p or Δp, g, and h are in lbs/ft^2, ft/sec^2, and ft, respectively.

Liquid density measurement can also be made with a bubbler system. As shown in Figure 11-6, air is introduced to two tubes immersed in the liquid. The lower ends of the tubes are at different elevations. Air pressure to each tube is regulated so as to be slightly greater than the pressure in the liquid. As the liquid density changes, the pressures at the ends of the tubes change. Density can be calculated from the following:

$$\rho = \frac{\Delta p}{g \Delta h} \quad (11\text{-}3)$$

where Δp is the difference in pressure supplied to the two tubes, g is the acceleration of gravity, and Δh is the difference is submersion for the tubes.

In measuring density by radiation, the density and thickness of the material under study, the pipe or container, and any insulation on the pipe or container are all important. Any changes in radiation are basically a result of differences in

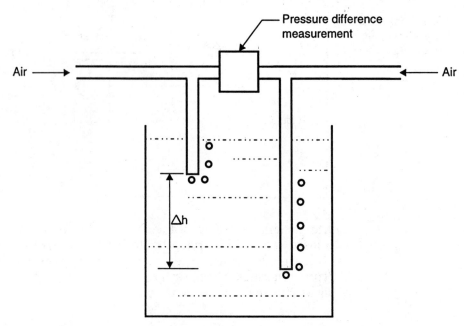

Figure 11-6 Set-up for measuring liquid density with bubbler system.

the density of the material under study. The radiation source is located on one side of the pipe or container and the sensing device on the other. Figure 11-7 shows a radiation density sensor.

Measurement of gas density is usually either by sensing the changes in frequency of oscillation of a vane immersed in the gas or by using a device which essentially weighs a volume of the gas and compares that weight to an equal volume of air.

Application Considerations

In general, there should be just enough agitation of the liquid under consideration to ensure uniformity of the density. Too much agitation could lead to velocity effects on the measurements.

Although not mentioned in the previous section on measuring devices, density-measuring equipment is available for extreme temperatures and pressures (i.e., from less than –100° F to more than 500° F and pressures in excess of 1,000 psi). For corrosive, abrasive, and other unusual applications wherein use of probes might not be feasible, radiation devices should be considered.

Figure 11-7 Radiation density sensor *(Courtesy of Ohmart Corporation)*

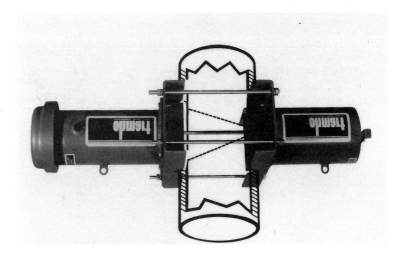

EXAMPLE 11-1

A liquid stands 18 inches deep in a rectangular tank measuring 2 feet by 4 feet. The liquid weighs 600 pounds. What are its density and specific gravity?

$$\text{Volume of liquid} = (2)(4)(1.5) = 12 \text{ ft}^3$$

$$\text{Specific weight of liquid} = \frac{600 \text{ lbs}}{12 \text{ ft}^3} = 50 \text{ lbs/ft}^3$$

$$\text{Density of liquid} = \frac{50 \text{ lbs/ft}^3}{32.2 \text{ ft/sec}^2} = 1.553 \text{ slugs/ft}^3$$

$$\text{Specific gravity} = \frac{50 \text{ lbs/ft}^3}{62.4 \text{ lbs/ft}^3} = 0.801$$

VISCOSITY

Some discussion of viscosity was included in Chapter 9. A portion of the information given in that chapter is included in the following discussion of viscosity and its measurement.

Basic Terms

There are forces existing between the molecules of a fluid which tend to cause the fluid to resist being deformed or set in motion. If a force is applied to a fluid at rest as shown in Figure 11-8, the fluid tends to move in layers with the relative movement as indicated in the figure. The property of the fluid which describes its ability to resist motion is called its *viscosity*. For many fluids, the force required to move the fluid is given by the following:

$$F = \frac{\mu A V}{y} \qquad (11\text{-}4)$$

where F is the force, A the boundary area being moved, V the velocity of the moving boundary, and y the distance between the boundaries. The Greek letter µ is the proportionality constant that takes into account the measuring units.

The force per unit area, F/A, is called the shear stress, τ. The equation for µ becomes:

$$\mu = \frac{\tau y}{V} \qquad (11\text{-}5)$$

µ is known variously as the viscosity, dynamic viscosity, or coefficient of viscosity. In the form of Equations 11-4 and 11-5, its units are lbs sec/ft². Various conversion factors for viscosity are given in Table 9-1. Because much of the information in tables and handbooks is given in terms of poise or centipoise (0.01 poise), it may be helpful to point out that water at 68.4° F

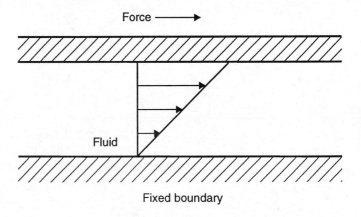

Figure 11-8 The fluid layers closest to the moving boundary move the fastest.

Table 11-1 Dynamic viscosities for some gasses and liquides at 68° F and standard atmospheric pressure (lb sec/ft^2)

Air	0.00000038
Carbon dioxide	0.00000031
Hydrogen	0.00000019
Nitrogen	0.00000037
Oxygen	0.00000042
Carbon tetrachloride	0.000020
Ethyl alcohol	0.000025
Glycerin	0.018
Mercury	0.000032
Water	0.000021

has a viscosity of 0.01 poise. Table 11-1 lists typical dynamic viscosities for some gases and liquids.

It is interesting to note that the viscosity of a liquid decreases as the temperature increases. This is due to a decrease in resistance to the motion of the fluid molecules and layers. For a gas, an increase in temperature causes an increase in the number of collisions between atoms or molecules. This leads to an increase in viscosity for an increase in temperature.

Before beginning the description of viscosity-measuring devices, it should be pointed out that not all fluids behave in the linear (straight line) manner depicted in Figure 11-8. Those that do are called *Newtonian* fluids. In general, the viscosity of a Newtonian fluid is constant for any given temperature. Some fluids, called non-Newtonian, have viscosities that vary with other parameters. Rate of flow is one of the parameters affecting some non-Newtonian fluids.

Measuring Devices

The devices to be described are known as *viscometers* or *viscosimeters*. It should be noted that they are designed to measure the viscosity of Newtonian fluids. The characterization of flow for non-Newtonian fluids is more complicated and beyond the intended scope of this book.

A sphere or a cylinder falling through the fluid of interest can be used to measure viscosity. With this method, it is necessary to accurately measure the time required for the sphere or cylinder to fall a given distance. Obviously, the higher the viscosity, the longer the travel time. A continuous measuring device can be set up using this method if a means for returning the sphere or cylinder to its starting place can be arranged. An inline falling-cylinder viscometer is shown in Figure 11-9.

In a drag-type viscometer, the space between two concentric cylinders is filled with the fluid of interest (see Figure 11-10). The inner cylinder is rotated by an electric motor while movement of the outer cylinder is restrained by springs installed so as to resist the torsional motion. The viscosity of the fluid causes a drag on the outer cylinder, and the amount of spring deflection can be related to the fluid's viscosity. Continuous readings can be made if means are provided for the liquid to enter and leave the device.

Viscosity can also be determined by measuring the time required for a given amount of the fluid to pass through an orifice of a given size. One of the devices using this principle

Figure 11-9 Inline viscometer
(Courtesy of Norcross Corporation)

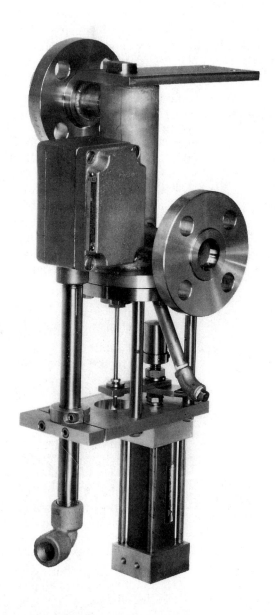

is called a Saybolt universal viscometer. Its reading is in Saybolt universal seconds, which can be converted to other viscosity units. In some cases, the orifice is replaced by a short section of pipe with a uniform, accurately measured inside diameter called a capillary tube.

Figure 11-10 Concentric cylinder viscometer.

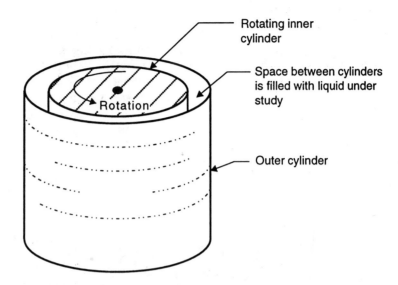

Although probably not as accurate as the aforementioned methods, it is possible to estimate fluid viscosity by timing the rise of air bubbles in the liquid of interest.

Calibration

Accurate measurement of fluid viscosity is not always an easy task. Care must be taken if the results are to be used for more than making estimates. In general, viscosity-measuring devices are calibrated using liquids of known viscosity or by comparing their results to those from other devices known to be giving accurate results.

EXAMPLE 11-2

A liquid with a viscosity of 5×10^{-5} lb sec/ft^2 is contained between two parallel plates located 0.6 inch apart. If the top plate is given a velocity of 10 feet per second parallel to the bottom plate, what is the force acting on each square foot of the moving plate?

$$F = \frac{5 \times 10^{-5} \text{ lb sec}}{\text{ft}^2} \left[\frac{1 \text{ ft}^2 \left(\frac{10 \text{ ft}}{\text{sec}}\right)}{0.5 \text{ ft}} \right] = 0.01 \text{ lb}$$

POSITION/DISPLACEMENT

Measurement of position or displacement is important for such applications as rolling mills, machining operations and numerically controlled tools. Also, some transducers for measuring variables such as temperature or pressure use position or displacement in converting the variable change to an electrical signal. Position transducers can be arranged to measure either linear or angular displacement.

Basic Terms

Position measurement can be either *absolute* or *incremental*. An absolute device constantly measures position with respect to some fixed reference point. An incremental device indicates the distance moved but is different than the absolute device in that it would not give a correct reading after power had been interrupted. Incremental measuring is usually cheaper, but a reference point must be designated at each start-up.

Measuring Devices

A potentiometer is probably the simplest displacement-measuring device. The wiper of the potentiometer is mechanically connected to the object whose displacement is to be sensed, and an electrical signal directly proportional to the wiper position is produced. Potentiometer devices can be used over a wide range of displacements, ranging from small fractions of an inch to several feet. They also can be designed to measure angular displacement. The chief disadvantage of this type of device is its relatively short life in some heavy-use applications. Another possible shortcoming is that a dirty spot on the track could cause the device to stop sending a signal when installed in a closed-loop system.

One of the best devices for measuring small displacements is the linear variable differential transformer (LVDT). The basic system consists of a transformer with two secondary windings and a movable core. When the core is in the center position, the voltage in the two secondary windings will be equal and the output zero. Movement of the core from the center position changes the voltages and indicates displacement. Among the advantages of LVDTs are that they can be electronically isolated and that they can be noncontact devices. Their expected life is much longer than the potentiometer devices.

Among the other devices that can be used for position or displacement measurement are torque transmitters and encoders. The torque transmitter senses an angular displacement through the torque applied to a shaft. The encoders measure angular displacement using binary notation.

FORCE, TORQUE, AND LOAD CELLS

Many industrial situations require knowledge of the "load" acting at a given location. The form of this load may be as a force or a moment (moment is sometimes referred to as torque). Also, many of these applications are related to the weight of a given amount of the material of interest. The devices used to measure the force or moment are often called *load cells*.

Basic Terms

Mass is a measure of the quantity of matter under consideration. The term *force* is a derived unit which relates the mass of an object with its acceleration through the following formula:

$$\text{force} = (\text{mass})(\text{acceleration})$$

Much of the time, the force of interest is the weight of the object or material under consideration, which then leads to the following:

$$\text{weight} = (\text{mass})(\text{acceleration of gravity})$$

Forces are defined by their magnitude (size) and direction (see Figure 11-11). In general, a force can be assumed to be acting anywhere along its line of action. When a force acts on a body, it tends to move the body along the force's line of action. Depending on where the force acts on the body, it may also tend to cause the body to rotate. The amount of rotation is dependent on the *moment* involved. By definition, the moment of a force is the product of the magnitude of the force and the perpendicular distance from the line of action of the force to the point being considered (see Figure 11-12). The resulting equation is as follows:

$$M = Fd \qquad (11\text{-}6)$$

where M is the moment in lb ft, F is the force in pounds, and

Figure 11-11 A force has magnitude and direction.

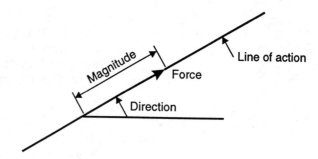

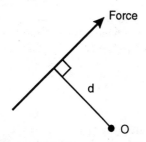

Figure 11-12 The moment of a force is found by multiplying the magnitude of the force by the perpendicular distance from the line of action of the force to the point in question.

d is the perpendicular distance between the body and the line of action of the force.

In some applications, this moment is called torque because it tends to cause a twisting action. In the special case shown in Figure 11-13, the moment caused by the two forces equal in magnitude but opposite in direction is called a *couple*.

Measuring Devices

There are two basic methods for measurement of force (including weight) measurement. Direct comparison uses some form of beam balance with a nulling procedure. Indirect comparison uses some sort of transducer.

Analytical Balance. The analytical balance is probably the simplest device for force measurement. It actually operates on the basis of moment comparison. From Figure 11-14, if arms L and R are the same length, the pointer P which pivots

Figure 11-13 Two forces equal in magnitude but acting in opposite directions form a couple.

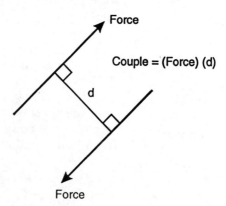

about 0, will be vertical when the weights at A and B are the same. By adding objects of known weight at A until the balance "balances," the weight at B can be determined. Note also that if arms L and R were not the same length, it would still be possible to determine an unknown weight at B by using the fact that the moments resulting from the weight at A and B about point 0 must be equal if overall balance is to be achieved. If L were two units long and R four units long, one could establish that, at balancing:

(4)(weight at B) = (2) (weight at A)

weight at B = ½ the weight at A

It is on the basis of this example that multiple-lever balance systems are designed to take care of various weighing problems.

EXAMPLE 11-3

The left side of a balance has a 6 inch arm and the right side has a 24 inch arm. What counterweight will be required on the right if a 3 ounce item is placed on the left side?

(3 ounces) (6 inches) = (? ounces) (24 inches)

number of ounces required = 0.75

Elastic Transducers. The simplest elastic-type transducer uses a spring as shown in Figure 11-15. For a simple linear spring, the following equation holds:

$$F = Kx \qquad (11-7)$$

Figure 11-14 Analytical balance operation.

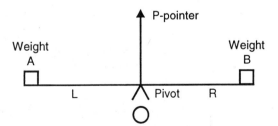

Figure 11-15 Elastic spring force (weight) transducer.

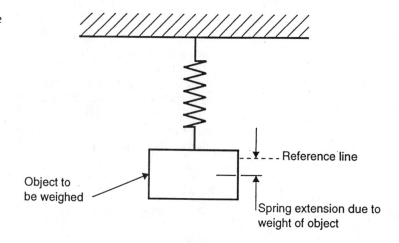

where F is the force or weight in pounds, K is the spring constant in lbs/in, and x is the spring deflection in inches.

EXAMPLE 11-4

A spring has an elongation constant of 50 lbs/in. When a certain object is hung from one end of the spring, it deflects 1.875 inches. What is the weight of the object?

$$\text{weight} = 50 \frac{\text{lbs}}{\text{in}} (1.875 \text{ in}) = 93.75 \text{ lbs}$$

Hydraulic and Pneumatic Systems. If a force is applied to one side of a piston and hydraulic or pneumatic pressure applied to the other side, a balance between the effects of the force and the pressure leads to measurement of the force. From Figure 11-16, the formula for measuring the force becomes as follows:

$$F = pA \tag{11-8}$$

where F is the force, p is the pressure behind the piston, and A is the piston area available for the pressure to act on. Note that this type of system is subject to friction between the piston and the wall of the cylinder and to leaks in the seals between the piston and walls.

Figure 11-16 The magnitude of the force is equal to the pressure multiplied by the cross sectional area of the piston (using consistent units).

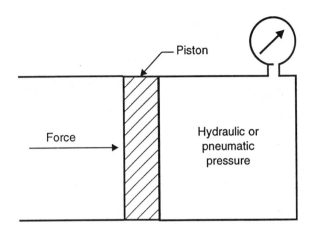

EXAMPLE 11-5

The piston in a hydraulic pressure force measurement device has a 10 inch diameter. If the hydraulic pressure gage reads 13.5 psi, what is the force acting?

$$\text{force} = 13.5 \frac{\text{lbs}}{\text{in}^2} \left(\frac{\pi \, 10^2 \, \text{in}^2}{4} \right) = 1{,}060 \text{ lbs}$$

Other Force-Measuring Methods. Other types of force-measuring systems include magnetoelastic devices which sense changes in magnetic permeability, piezo-type cells which use either piezoresistivity or charge-producing principles, and linear variable displacement transformers where the transformer measures the deflection of a spring.

Dynamometers. Torque associated with mechanical power is usually measured with a *dynamometer*. These devices generally fall in one of three categories: mechanical, hydraulic, or electrical. They are used for such applications as measuring power or torque developed by electric motors or engines, supplying specific amounts of energy to test devices, or simply to monitor torque at a particular location.

Load Cells. In the broadest sense, the term *load cell* is used to designate the device for measuring force or torque. In addition to the various principles noted earlier, loads can be sensed with devices that incorporate electric strain gages. The load to be measured causes a small change in the geometry

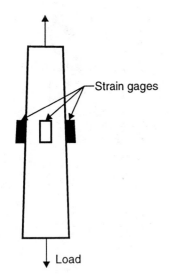

Figure 11-17 Strain gages can be used to measure load.

of the device on which the strain gages are attached, and thus the output of the gages is changed. The output is then related to the load.

As an example, strain gages could be attached to a column as shown in Figure 11-17. Because the strain in a column can be calculated from engineering mechanics equations, the load corresponding to a particular strain can be calculated. In addition to a column in tension, the strain gage could be attached to members in compression, those acting as cantilever beams, and for some special applications such as proving rings.

Application Considerations

In some applications, care must be taken to allow for temperature effects. These effects can be due to modulus of elasticity changes or to changes in geometry. Some devices, particularly the electric strain gaged ones, can be designed to minimize or eliminate the temperature effect.

Readings from some measuring systems are subject to system vibration. This is most easily visualized in the case of the spring. It is possible to reduce this effect through the use of vibration dampers. This is not always a straightforward exercise because it is possible to overdamp a system (in which case the true force is never reached) or underdamp (in which case there is still too much vibration).

SOUND

There are two essentially different reasons for being interested in the understanding and measurement of sound. The first reason relates to the human sense of hearing. The second reason includes both structural failure that can be due to sound excitation and the use of sound to study and deal with deforming and fracture processes.

The term *noise* is usually taken to mean an unwanted or undesirable sound. Noise can affect human activities in either a positive or negative way. Certain sounds and sound levels can be soothing to the human ear and promote well-being and can help set the environment for fruitful working conditions. On the other hand, other sounds and sound levels can interfere with human communication, reduce worker efficiency, and cause permanent hearing damage. It is not always easy to separate wanted and unwanted and useful and harmful effects of sound.

Basic Terms

The basic form of sound is a single frequency. This is sometimes referred to as *pure tone* and is difficult to produce because of reflections that are present. *Random noise* is a much more common subject. It is comprised of sound coming from more than one source. Sound reaching an object usually comes from more sources than the source originally producing the sound (*primary source*). *Secondary sources* are any objects in the vicinity which receive the sound from the primary source and then reflect part of it (the rest is absorbed). Figure 11-18 shows part of the activity that takes place between a primary source and a receiver.

The combining of sounds from primary and secondary sources produces *standing waves*. These standing waves cause the sound at one location to be "louder" or "quieter" than at another even though the location of the sources would lead one to think the sound level distribution should be otherwise.

The *sound pressure level* (SPL) is the difference between the air pressure at a point at a given instant in time and the average air pressure at that point. The units involved are those of pressure. Those commonly used include the following:

$$1 \text{ dyne/cm}^2 = 1 \text{ }\mu\text{bar} = 1.45 \times 10^{-5} \text{ psi} \tag{11-8}$$

The term *decibel* (dB) is used to compare different sound levels. It is taken as the following:

$$\text{number of dB} = 20 \log_{10} \left[\frac{\text{pressure}_1}{\text{pressure}_2} \right] \tag{11-9}$$

In this form, the decibel is a quantity relating the sound level at two locations. A pressure of 0.00002 N/m² is the accepted reference with which other pressures are compared. If this

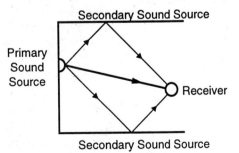

Figure 11-18 Simple example of primary and secondary sound sources.

value is placed in the denominator in Equation 11-9, the number of dB resulting is in terms of those most frequently quoted. This form of comparison is used because it represents the average threshold for human hearing at a frequency of 1,000 Hz. Note that the 2 dB is approximately 3×10^{-9} psi. Examples of typical dB ranges are the following: onset of pain for humans is approximately equal to 140–150; rocket engines is approximately equal to 170–180; and a factory is approximately equal to 80–100.

Sound energy can be expressed in terms of power. The intensity at a location is given as watts per unit area. The sound pressure level resulting from the combination of more than one sound is not the algebraic sum of the individual sound levels. An equation that can be used to obtain the pressure level for two sounds is as follows:

$$SPL = 10 \log \left[\log^{-1} \frac{dB_1}{10} + \log^{-1} \frac{dB_2}{10} \right] \quad (11\text{-}10)$$

where dB_1 and dB_2 are the sound levels for the two sounds being combined.

As sound moves radially through space from a source, the sound level decreases according to the following:

$$\text{change in sound level} = 20 \log \frac{r_b}{r_a} \quad (11\text{-}11)$$

where r_b and r_a are the distances in space from the two points under consideration and the sound source.

Two other terms should be included at this point. *Loudness* is a subjective quality and measures relative sound strength. It has a linear unit called *sone*. *Loudness levels* define the difference in loudness and are measured in *phons*. Since almost all sounds are made up of several frequencies (tones), it is often necessary to perform a *frequency spectrum analysis*. This process takes the complex sounds and sorts them into various frequencies, ranges, or bands. It is often useful for finding the major source of a sound problem.

Measuring Devices

Sound-measurement systems usually include (1) a microphone for picking up the sound energy and transferring it into another form of energy, (2) modifying devices which include amplifiers and filters, and (3) meters or recorders. Most microphones have a thin diaphragm—which senses the sound—and transducers—which may be capacitors, crystals, or electronic devices.

The sound level meter is probably the most commonly used device for routine measurement. Many varieties are available. Some are all-purpose devices, while others are designed to cover only certain ranges and situations. A more specialized device called a spectrum analyzer provides direct information on the various sources comprising a given sound. An active-band sound analyzer is shown in Figure 11-19.

Application Considerations

Selection of a microphone depends on the ultimate use of the information but can include characteristics such as nondirectivity, uniform response over desired range, output much larger than the system's inherent noise level, and insensitivity to all but the sound under study.

Whether to aim the microphone directly at the sound source depends on the particular study. It may be necessary to consult the manufacturer or supplier for information in this regard. Some advise an angle of 65 to 75 degrees from the source. The best distance for the microphone to be separated from the sound source is in the area (a) far enough from the source to reduce the effect of other sources operating in conjunction with the source under study and (b) not so far away as to begin to pick up sound from other nonrelated sources.

Calibration of sound-measuring equipment is usually with an internal circuit supplied by the manufacturer or by comparing results with a standard.

EXAMPLE 11-6

What is the change in sound level at two points located in line with a source if the nearer point is 50 feet from the source and the points are 25 feet apart?

From Equation 11-11:

$$\text{change} = 20 \log \frac{50}{75} = -3.52$$

Figure 11-19 Active-band sound analyzer *(Courtesy of QuadTech Inc.)*

pH MEASUREMENTS

Basic Terms

There are many situations in industrial activities when water is used as an important part of a process. Quite often it is necessary to control the pH of the aqueous (aqueous means "containing water") solution. The term *pH* comes from "power of hydrogen" and is used to indicate the activity of the hydrogen ions in a solution. Another way of putting it is that pH describes the acidity or alkalinity of the solutions.

In a liquid such as pure water, the pH value is 7. This would be considered a neutral reading. When the number of hydrogen ions is increased, the solution becomes more acidic and pH value decreases. On the other hand, as the number of hydrogen ions decreases (and the number of hydroxyl ions increases), the pH value increases and the solution becomes more alkaline. A change of one pH unit means that the hydrogen ion concentration has changed by a factor of 10. For example, a pH of 1 indicates 0.1 gram/liter hydrogen ions; and a pH of 2 indicates 0.01 gram/liter.

Another way to consider the pH number is to look at the formula defining it:

$$pH = \log_{10} \left[\frac{1}{\text{hydrogen ion concentration, g/l}} \right] \quad (11\text{-}12)$$

A pH value can range from 0 to 14. Examples of highly acidic solutions are hydrochloric and sulfuric acids (pH around 0 to 1), while highly alkaline solutions include caustic soda (4% caustic soda pH = 14). Other pHs include lemon and orange juice (about 2 to 3) and ammonia (about 11).

Measuring Devices

There are two general methods used to measure the pH of a solution: chemical indicators and pH meters. Chemical indicators change color with a change in hydrogen ion concentration. In general, their best accuracy is approximately 0.1 pH unit. Pink and blue litmus paper can be used for general indication as to whether a solution is acidic, neutral, or alkaline. If after being immersed in a solution, pink litmus turns

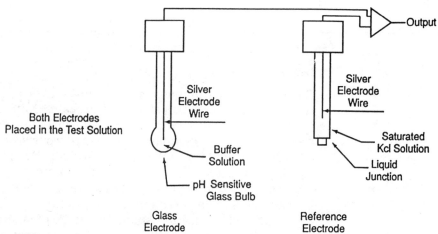

Figure 11-20 A pH cell for measurement of acidity or alkalinity of solutions *(From Jack W. Chaplin,* Instrumentation and Automation for Manufacturing, *© 1992, Delmar Publishers, Inc., Albany, N.Y.)*

white, the solution is either acidic or neutral; if it turns to a brighter pink, a strong acid is present. The same type of comments hold true for blue litmus paper, except the term alkaline is substituted for acidic.

A pH meter contains sensing devices and the necessary electronic and readout equipment. The sensing devices are usually a measuring electrode and a reference electrode (see Figure 11-20). These two electrodes form an electrolytic cell whose output is then related to the solution pH. Figure 11-21 shows a typical glass electrode.

Application Considerations

It should be noted that pH varies with temperature. The pH of 7 quoted earlier for pure water is for a temperature of 77° F (25° C). As the pure water temperature approaches 32° F, the pH approaches 7.5; as it approaches 212° F, the pH approaches 6. The pH measurement systems usually include built-in temperature compensation.

In process operations, where solid and liquid material may be being added or removed, location of pH-measurement devices and their readings must be done with care. Dynamically changing conditions can change pH quickly. Attention should be given to keeping probes clean.

Calibration of pH equipment is usually carried out with commercially available buffer solutions with known pH values. When moving from one solution to another, the sensors should be rinsed carefully so as not to cause a calibration error.

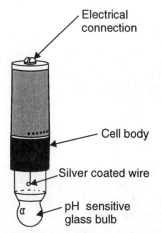

Figure 11-21 Glass-bulb pH electrode

EXAMPLE 11-7

How much does the pH of a solution change if the hydrogen ion content goes from 0.03 g/L to 0.001 g/L?

$$pH_1 = \log_{10}\left[\frac{1}{0.03}\right] = 1.523$$

$$pH_2 = \log_{10}\left[\frac{1}{0.001}\right] = 3$$

change in pH = 1.523 − 3 = −1.477

REVIEW MATERIALS

Important Terms

density
specific weight
specific gravity—gas
specific gravity
hydrometer
viscosity
shear stress
poise
Newtonian fluid
non-Newtonian fluid
viscometer or viscosimeter
absolute position
incremental position
force
moment
torque
loads cells
couple
direct-comparison force measurement
indirect-comparison force measurement
analytical balance

dynamometer
noise
pure tone
random noise
primary source
secondary source
standing waves
sound pressure level
decibel
loudness
sone
loudness level
phons
frequency spectrum analysis
pH
aqueous solution
acidic solution
alkaline solution
chemical indicators
litmus paper
pH meter

Questions

1. Discuss the difference between density and specific gravity.
2. Discuss the difference in the definitions of specific gravity for liquids and gases.
3. How is the phenomenon of vibration used in determining the density of a liquid?
4. What are some of the items to be considered when designing a bubbler system for measuring the specific gravity of a liquid?
5. Discuss the difference between a Newtonian and non-Newtonian fluid.
6. Why does the viscosity of a gas increase as the gas temperature increases?
7. Discuss the possible effects of sphere or cylinder size, container inside diameter, and length on the falling sphere or cylinder viscometer.
8. Describe the difference between absolute and incremental position measurement.
9. How does a potentiometer displacement-measuring device work?
10. Which position/displacement-measuring device discussed is usually best for making accurate small displacement readings?
11. What are the advantages of the linear variable differential transformer devices?
12. Discuss the difference between force and torque.
13. Why is a "cheater" sometimes used to move a large object (a cheater is an extension of a pry bar, wrench, etc.)?
14. Discuss the differences between using direct and indirect methods for measuring force or moment.
15. What are the advantages and disadvantages of using hydraulic or pneumatic systems for force measurement?
16. Discuss the two essentially different reasons for measurement of sound.
17. What is a pure tone?
18. What is random noise?
19. What are standing waves, and why are they important?
20. Discuss a term that is used to compare different sound levels.
21. Describe the makeup and discuss the application of a microphone.
22. What does a pH of a solution indicate?
23. The pH of a solution in a process is to be accurately measured. The solution temperature will vary from 50° F to 90° F. Discuss whether temperature compensation is necessary for the pH equipment.
24. If a process involves the use of nitric acid, would you expect the pH to be above or below 7?
25. The pH of a solution was measured at 180° F. No changes were made except to lower the temperature to 100° F. Would you expect a change in pH and, if so, in which direction?
26. A co-worker tells you that, using litmus paper, he found the pH of a solution to be 9.31. What is your comment?

Problems

1. The difference in air pressure supplied to the two tubes in a bubbler system is 0.44 psi. If the vertical distance between the bottom ends of the tubes is 24 inches, what is the liquid density?
2. A constant-level tank (as in Figure 11-5) contains a liquid with a density of 1.5 slugs/ft^3. If the pressure at the base is 11.0 psi, what is the distance between the constant level and the bottom of the tank?
3. What is the density of water having specific weight of 62.4 lbs/ft^3?
4. If the density of water on earth is 1.938 slugs/ft^3, what would its specific weight be on a planet with an acceleration of gravity of 5.2 ft/sec^2?
5. A certain material has a specific weight of 327 lbs/ft^3. What is its specific gravity?
6. Taking the specific weight of air under standard conditions as 0.076 lb/ft^3, what would be the specific gravity of a gas with a density of 0.0041 slug/ft^3?
7. If the force measured on an area as indicated in Figure 11-22 is 0.2 lb, what is the fluid's viscosity? The distance between the walls is ¼ inch, and the 0.25 ft^2 area is moving at 1.5 ft/sec.

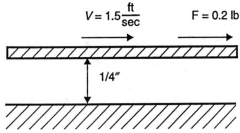

Figure 11-22 Figure for use with Problem 7.

8. Two plates separated by 2 inches are placed in a fluid with a viscosity of 133 x 10^{-6} lb sec/ft^2. If the plates areas are 4 ft^2, what velocity would one of the plates reach if a force of 0.1 lb is applied to one of the plates in a direction parallel to the plates. The second plate remains stationary.
9. Recalling that kinematic viscosity is given by $v = \mu/\rho$, what is v the kinematic viscosity of a fluid with density of 1.8 slugs/ft^3 and a dynamic viscosity of 6 x 10^{-4} lb sec/ft^2?
10. If the gap between the cylinders in a drag viscometer is reduced by 50%, what effect would it have on the torque reading of the stationary cylinder?
11. What shear stress exists on a plate moving at 2 ft/sec and separated from a stationary plate by ¼ inch? The viscosity of the fluid between the plates is 1.042 x 10^{-4} lb sec/ft^2.
12. How fast would a 10 ft^2 plate located ½ inch from a horizontal wall be moving if the force causing it to move was 0.4 lb? The viscosity of the fluid is 1.66 x 10^{-3} lb sec/ft^2.
13. If proper units are being used, what weight is associated with a mass of 2.0 and an acceleration of gravity of 32.2?

14. To obtain a moment of 10 lb ft, what force must be used if the moment arm is 2 feet? If the moment arm is 6 inches?

15. In Figure 11-23, with no force at B, what force at A is required for balance? If that force is removed, what force would be required at B for balance? What was the initial force at A?

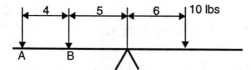

Figure 11-23 Figure for use with Problem 15.

16. Two forces act as shown in Figure 11-24. What is the couple they create? If the distance between the forces was reduced to 3½", what would the couple become?

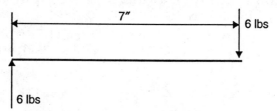

Figure 11-24 Figure for use with Problem 16.

17. For the analytical balance shown in Figure 11-25, what would be the relationship between the weight on the right and the one the left?

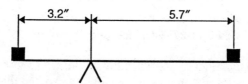

Figure 11-25 Figure for use with Problem 17.

18. A 1.5 lb weight hanging from a spring increases the length of the spring to 8 inches. If the spring constant is 2.4 lbs/in, what is the unloaded length of the spring?

19. If the sound pressure from one source is 0.00006 N/m², what is the dB reading?

20. What is the dB reading associated with a measured pressure of 0.00002 N/m²? Why is it significant?

21. What pressure would correspond to a dB reading of 50?

22. What sound pressure would exist in a factory with a dB of 90?

23. What is the sound pressure level for two machines, each with a dB level of 100?

24. Two machines are located 200 feet and 100 feet from a sound source. How much does the sound level decrease from the near to the far point?
25. How many grams per liter hydrogen ions are present if the pH is 3? If the pH is 1?
26. 0.0005 gram per liter hydrogen ions are present in a solution. What is the pH reading?
27. A solution has a pH reading of 10.7. What is the hydrogen ion concentration?
28. What is the difference in hydrogen ion concentration for two solutions, one having a pH of 2 and the other with a pH of 4?
29. If the pH of a solution changes by 3, by what factor does the hydrogen ion concentration change?
30. The hydrogen ion content of a solution is 1.5×10^{-6} gram per liter. Is this an acid or alkaline solution?

Process Control

CHAPTER GOALS

After completing study of this chapter, you should be able to do the following

- Define the meaning of controller action.
- Describe on/off action and draw a graphical sketch for any on/off system.
- Explain the concept of lag time.
- Understand the meaning of error signal and correction signal.
- Draw a proportional band for a process control system, given the amplification factor and the actuator control factor.
- State exactly how derivative action works and for what purpose it is used.
- Discuss the consequences of too much and too little amplification in proportional control.
- Explain why integral action was developed and give an example of how it works.
- Know the difference between a long-term load and a short term fluctuation in the measured variable signal.
- Define the meaning of the acronym PID and describe why it is the best known of all controller action modes.
- Understand the workings of a relatively simple pneumatic proportional amplifier.
- Describe in general the two most popular digital controller systems.

The typical industrial plant is saturated with process control systems. The heating and air conditioning of the plant's working space is a controlled process, albeit a simple one in most cases. The regulation of the amount of each ingredient that flows into a mixture of gunpowder is a controlled process—a very highly controlled and closely regulated process. Anyone that works as a technologist or engineer in an industrial setting needs to know at least the basics of process control because, even if that person is not directly responsible for any of the control systems, he or she is surrounded by them. Regardless of job function, sooner or later any industrial employee is going to have to be concerned with the workings of a process control set-up.

The centerpiece of process control is the controller. The main topics of this chapter are the types of controllers that are available and the various modes of control action. There are many different types of controllers, including the following and combinations of the following:

- Mechanical
- Pneumatic
- Hydraulic
- Electromagnetic
- Analog electronic
- Digital electronic

There are also many different means by which controllers can respond to changes in the measured variable. These means are known as *controller action* and include such actions as on/off and proportional. This textbook addresses the more common types of controller action and then discusses how the three principal types of controllers (analog electronic, pneumatic, and digital electronic) carry out these actions.

CONTROLLER ACTION

On/Off Control Action

The simplest control action is the on/off system. *On/off control action*, also known as *two-position action*, is a process system whereby the actuator has only two settings (on and off), and the controller turns the systems on (or off) whenever the measured variable is above the set point and off (or on) whenever the measured variable is below the set point. At first, one might think that the system would be continually cycling on and off rather rapidly, with very little pause between the on stage and the off stage. This problem usually does not arise, however, because of lag time. *Lag time* is the time it takes a control system to return the measured variable to its set point after the measured variable has strayed from its set point. An example of simple on/off action is shown in Figure 12-1.

In Figure 12-1, time is represented left to right on both the top and bottom graph. This is the common procedure used on all graphs in this chapter that are used to show how a particular controller action works. Usually the horizontal time axis is not labeled in regard to exactly the amount of time because it is the concept that is being presented, not an exact case. Always, however, the horizontal axis represents increasing time, left to right; and all graphs that are in the same figure but that are arranged either above or below one another are on exactly the same time scale.

The top graph in Figure 12-1 represents the measured variable—in this case, the room temperature. The bottom graph represents the action taken by the actuator or, more exact, the percentage of the correction signal sent to the actuator by the controller. (Note: The correction signal is also called the manipulated variable signal.) This system would be a typical simple system used for air conditioning of a room or building. When the measured variable is below the set point of 72° F, the actuator is off. In other words, the relay that forwards electrical power to the cooling compressor and the fan is in

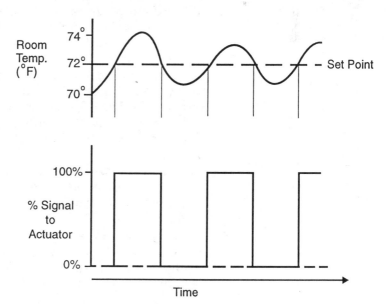

Figure 12-1 Simple on/off action being used for an air conditioning system.

its "off" position. This makes sense because the temperature is below the preferred temperature of 72° F and one would not want the air conditioning system to be running.

Since the system is in the "off" position, the temperature will finally rise until it reaches the set point. At that time, the controller will send a 100% correction signal to the actuator, and the actuator will turn the air conditioning system on. The temperature will now soon begin to drop, but not necessarily right away. Depending on the size of the room or building, the temperature might continue to rise for a few more minutes before the air conditioning begins to take effect and starts lowering the temperature. Unless the air conditioning system is simply too small to handle the job, the temperature will eventually be lowered to the set point and the system shut off. The time difference between when the system begins to try to lower the temperature to the set point and the time when it finally does lower the temperature to the set point is called the lag time, as mentioned at the start of this section. For the process indicated by Figure 12-1, the lag time would be the average time between the points where the measured variable crosses the set point line.

Differential On/Off Control Action

Differential on/off control action represents a mode whereby the actuator does not go into "on" or "off" mode until the measured variable is a small percentage either above or below the set point. A very good example of this is the ordinary home air conditioning system. If the thermostat is set at 72° F, the home air conditioning does not ordinarily cycle on until the air temperature at the thermostat is above the set point, say, at 74° F. Likewise, the actuator does not go into its 100% off position until the temperature at the thermostat has dropped to a temperature lower than the set point, perhaps 70° F.

An industrial example is the maintenance of level in a bin or container between a high point and a low point. Differential on/off control action is often used to prevent the rapid cycling on and off of the actuator. For example, if a water tank was relatively small and the input water supply flow much greater than the constant outflow, the input water supply would be continually switching on and off with only a few seconds between cycles if simple on/off action were being used.

EXAMPLE 12-1

Draw a diagram similar to Figure 12-1 representing a home central heating system using a differential on/off system with a ±3° F differential.

The diagram is shown in Figure 12-2. Notice that the actuator on/off action is just the opposite of that for an air conditioning system. Also, notice that the actuator does not go into the "on" stage until the measured variable reaches 75° F and that it does not go into the "off" stage until it reaches 69° F.

Figure 12-2 Differential on/off action being used for central heating.

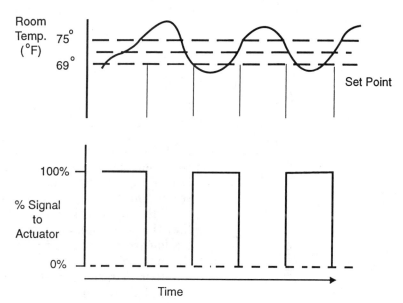

Proportional Action

The most common of all industrial process control actions is *proportional action*. Proportional action means that the percentage change of the manipulated variable signal (better known as the *correction signal*) sent by the controller is proportional to the percentage change of the measured variable signal that is transmitted to the controller. In other words, the controller is either amplifying or reducing the difference of the measured variable signal from the set point and sending it onward to the actuator. Figure 12-3 shows an example of a change in load, the resulting change in the measured variable, and the resulting output of the manipulated variable signal to the actuator.

Figure 12-3 A simplified example of how ideal proportional control action would respond to a load change when the change in the measured variable is negative if the change in the load is positive. The time represented here is very small.

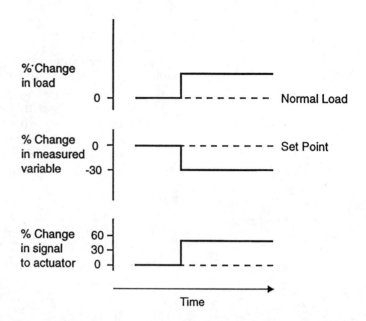

The *load* represents the demand being placed on the process control system for a needed amount of controlled output. For example, suppose a process control system regulated the water temperature for a hot-water cleaning station on a product line. Further assume that the product is a large one and that the time intervals between the product arriving at the station are relatively large, say, 10 minutes. Between the end of the cleaning of one product and the arrival of the next product, there is no load on the system, that is, no demand for the controlled item, the hot water. When the next item arrives, it is cleaned for 5 minutes and there is a load placed on the process system for 5 minutes. The vertical series of graphs in Figure 12-3 represent how the system responds during the first few seconds to the load each time. Please note that the percentage change in all the signals instead of the actual signal is plotted. By using this method, the graphs do not have to be modified for different units and different values in regard to the vertical axis. This provides for easy comparison as graph after graph is discussed in this chapter.

When cleaning starts (this is when the load rises from zero), the measured variable is going to change. In this case, since the measured variable is the temperature of the water, its value is going to drop, as indicated in Figure 12-3. (Actually, just because the temperature drops, the signal to the controller does not have to drop. The sensor/transducer might have signal conditioning built in that not only amplifies the signal but inverts it. However, to better conform with the situation as it is being described herein, the measured variable will be shown as going in the opposite direction of the load for this case. This is a very common scenario, and many texts always show the measured variable going in the opposite direction of the load. This convention does make the analysis of the ongoing action more complex, however; and this textbook, in general, uses the rule that the measured variable moves in the same direction—positive or negative—as the load. An exception is being made for this one example.)

Next, the proportional controller compares the measured value signal to the set point, amplifies the difference (known as the *error signal*), and sends a correction signal to the actuator. In Figure 12-3, the controller amplified a negative 30% change in the measured variable by the factor of 2 and inverted the result to obtain the correction signal. This indicates that this proportional controller had an amplification factor or *gain* of 2. However, the inversion and amplification of the controller cannot be set up as simply as this example illustrates. This is because the *error* (the difference between the measured variable and the set point) that the controller measures can theoretically range from $-\infty$% to $+\infty$%. This is not a realistic range, of course; and it is common to the use the conceptual error range of ±50%. On the other hand, the actuator is usually going to be set at its normal position of halfway open (or halfway on, etc.) so that it can be varied the maximum amount either way from this position. The result is that proportional controllers are usually designed so that the relationship between the percentage error input and the correction signal output (known as the *proportional band*) is as shown in Figure 12-4.

As mentioned in an earlier paragraph in this section, all subsequent controller action diagrams show the measured variable moving in the same direction as the load (i.e., the

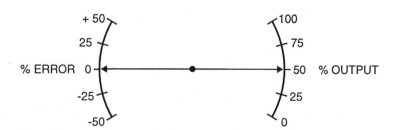

Figure 12-4 A typical example of *proportional band*—the relationship between the error input and the correction signal output of a proportional controller.

measured variable increases as the load increases) for convenience of analysis, even though in some cases the measured variable moves in the opposite direction. Figure 12-5 shows Figure 12-3 redrawn to represent this convention.

The first thing that might come to one's mind when viewing either Figure 12-5 or Figure 12-3 is that the controller action is having no effect on the measured variable. This is because both figures were intended to merely acquaint the reader with controller action diagrams. Figure 12-6 shows that the controller action should indeed have an effect on the measured variable.

Notice that the change in load causes a 30% increase in the size of the measured variable signal. This causes the controller to output a correction signal to the actuator that opens it to exactly the right setting needed to rapidly return the measured variable to its set point. Unfortunately, Figure 12-6 still does not represent a realistic process control result.

One problem with Figure 12-6 is that it assumes the controller has determined the exact amount of correction needed to bring the measured variable promptly back to its set point. When the controller corrects the measured variable enough so that it begins to approach the set point, the actuator also begins to return to its original (50% in the example) setting. This allows the measured variable to once again begin to swing

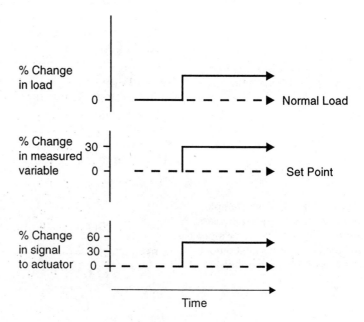

Figure 12-5 A simplified example of how ideal proportional control action would respond to a load change when the change in the measured variable is positive if the change in the load is positive.

Figure 12-6 An ideal example of how the action of the controller system should result in the measured variable being returned to the set point. This cannot actually occur for a simple proportional control system because a zero-error signal results in the actuator beginning to return to its original setting. This example is very closely achieved for short-term load settings, however.

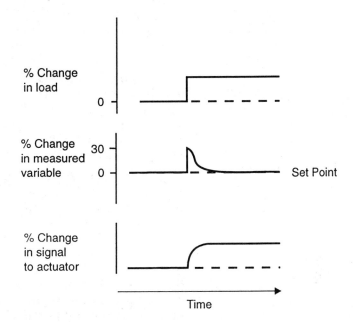

away from the set point, and the correction process begins once again. If the proportional band is set at a reasonable value, the system finally reaches a steady-state oscillation, as shown in Figure 12-7.

Figure 12-7 A more realistic example of how proportional controller action attempts to return the measured variable to the set point in long-term load situations.

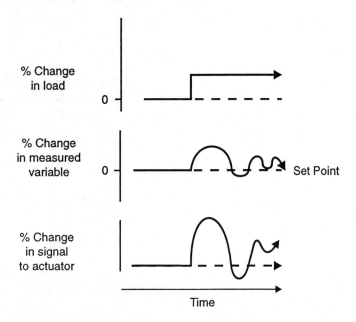

This type of control action is acceptable for many processes, especially those with only short-term changes. For those processes needing more accurate control, particularly those with long-term load changes, the concept of proportional control with integral reset action was developed. For those processes needing both accurate and fast action, the theory and procedure for proportional control with integral reset and derivative action was developed. The text covers the concept of derivative control first, then integral control, and then all three control processes together.

Proportional and Derivative (PD) Action

In an attempt to reduce correction time, a type of action known as *derivative action* has been developed. (It is also known as *rate action* and as *anticipatory action*.) This type of action is proportional to the *rate* with which the measured variable is changing, not with the amount of change. The rate of change at any given instant is defined in mathematics as the *slope* of the line that represents the plot of the dependent variable (which, in this case, is the measured variable) against the independent variable (which, in this case, is the time) at that same instant of time. Figure 12-8 shows plots of the value of the slopes of a couple of measured variable graphs immediately below the corresponding measured variable graph.

The purpose of Figure 12-8a is to reacquaint the reader with the basics of the algebraic meaning of the term *slope*. Among other things, the slope of a 45° line headed up is +1; the slope of a 45° headed down is –1; the slope of a horizontal line is 0; and the slope of line headed straight upward (as measured left to right in regard to time passing) is undefined or, as technological personnel more commonly phrase it, "positively infinite." The slope of a line headed downward is negatively infinite. The name derivative action comes from the calculus definition of slope at an instant of time.

EXAMPLE 12-2

Cover the slope graph portion of Figure 12-8b and sketch the slope of the percentage change of the measured variable in that figure.

Figure 12-8b represents the slope of a more common proportional curve. The very first thing the slope does is jump from

Figure 12-8 Simultaneous graphical examples of the measured variable and the corresponding slope of the plot of the measured variable.

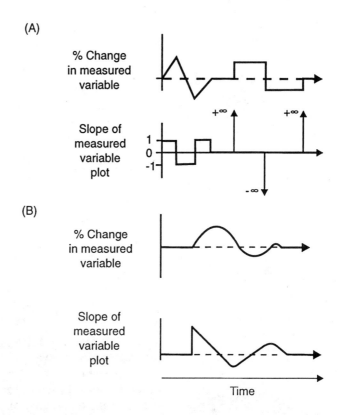

zero to a large value, then start downward. If this second curve is added to the first curve, then overall correction of the error in the measured variable takes place more rapidly. In other words, the slope of the top curve is added to the top curve, and the resulting sum represents the correction signal to the actuator. Because of this, the latter portion of the top curve in Figure 12-8b is incorrect. It does not reflect the effect that the contribution of the bottom curve has on error correction. A more accurate rendition of Figure 12-8b in actual practice is shown in Figure 12-9, where a short-term fluctuation in load has occurred.

Notice that by adding the value of the slope (the third graph from the top in Figure 12-9, known as the derivative action) to the change in the measured variable (the second graph from the top, known as the proportional action), a more effective total correction signal is sent to the actuator. It can be said that it is more effective because if one views the mea-

Figure 12-9 A graphical example of PD action.

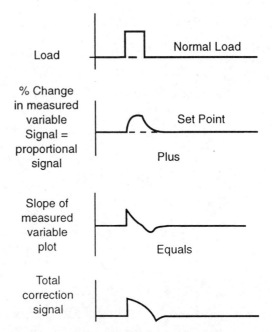

sured variable graph, it can be seen that the measured variable was much more quickly brought back to the set point than was the case with proportional action alone. Derivative action was developed to reduce lag time—to more rapidly return the measured variable to its set point. Proportional and derivative action (PD action) is very effective for short-term load fluctuations.

Proportional and Integral (PI) Action

The main purpose of *integral action* is to correct for long-term loads. The proportional controller alone cannot correct for long-term load changes but can only approximately correct for them, as shown in Figure 12-7.

As an example of long-term load demand, consider a hot-water supply system that typically supplies hot water for ten production lines. Because there are numerous on and off needs for hot water by one station on each of the ten production lines, the measured variable is continually moving above and below the set point and is probably easily being taken care of by correction signals sent out by the proportional controller with derivative action. About one day per week, however, the company produces a product that uses the same ten production lines but requires the use of two hot-water stations on each line. For the case of the system's controller, it now needs an average of 80% as its amplification midpoint instead of the usual 50%. The controller cannot maintain the actuator at this position, however, because as soon as the measured variable starts to approach the set point, the actuator signal starts back to its 50% position. The system does not work very accurately. For this reason, integral action was added to the control system.

Integral action, also known as *reset* action, is the computation of the area under the proportional action signal or, more exact, the computation of the area under the measured variable signal as it moves above and below the set point. The term *integral* is used because this is the calculus method of calculating area under curves. As an example, consider Figure 12-10.

Figure 12-10 gives some simple results of integration of a curve or, in simpler words, calculation of the area under a curve. The best way for the student to calculate the area un-

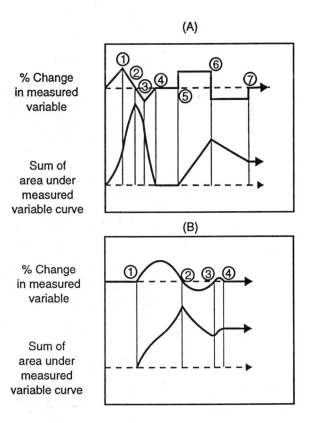

Figure 12-10 Simultaneous graphical examples of the measured variable and the corresponding sum of the area under the plot of the measured variable.

der the measured variable curve is to count the blocks of the graph paper under the curve. (One could also use geometrical formulas for calculating the area of known shapes, or a student who is familiar with calculus could actually integrate the curve. Even so, the most straightforward method is to count the blocks using a plot that is on graph paper.)

From the start of the measured variable plot in Figure 12-10a to point 1, the area under the curve rises rapidly. By counting the imaginary blocks, one can see this. At point 1, the area under the top curve is still increasing, but not very rapidly. At point 3, the blocks that are counted are under the axis and, hence, are negative. Thus, from point 2 to point 3, the total area under the curve decreases rapidly. The total area under the curve still decreases (but not as rapidly) from point 3 to point 4. At point 4, the sum of the area under the measured variable curve is zero because, to this point, the number of blocks under the curve above the axis is exactly equal to the number of blocks under the axis, between the curve and the axis.

From point 4 to point 5 there are no additional blocks under the curve; hence, the total area remains at zero. From point 5 to point 6, the area under the curve (the sum of the number of blocks) increases at a constant rate. From point 6 to point 7, the area decreases at a constant but lower rate. The net amount of area under the measured variable curve at point 7 is slightly positive. Since the curve is on the horizontal axis from point 7 onward, the area remains constant at the slightly positive value just mentioned.

EXAMPLE 12-3

First, cover the bottom graph of Figure 12-10b and then repeat the above procedure. Check the results with the bottom portion of that figure.

Neither of the examples in Figure 12-10 necessarily represents a long-term load because the load plot is not shown and because no units of time are displayed. What constitutes a long-term load? There is no definite answer. For some processes, 5 minutes of a sizable, somewhat constant, load increase could be considered "long term." For others, it might be 30 minutes or it might be 5 hours. As an example, suppose that 30 minutes of a new load is considered to be long term for a certain industrial process. The integration action portion of the controller would have its time constant adjustment set so that practically no integration will have taken place after 5 minutes. The signal going to the actuator might be 25% derivative (which is going away rapidly), 70% proportional, and 5% integral. After 15 minutes, there is a tiny bit (say, 15%) of integration signal contributing to the overall correction signal going to the actuator. At 25 minutes, 80% of the signal going to the actuator would be from the integration controller, and 20% proportional signal. At 30 minutes, the correction signal should be 100% integral. This process is graphically shown in Figure 12-11. The student should analyze this figure thoroughly, using the following as a guide:

1. Note that the load is long term by the definition of "long term" for this example.
2. Note that the derivative signal immediately surged upward and then quickly died away. Of course, by the definition of "slope" or "derivative," it should have gone to

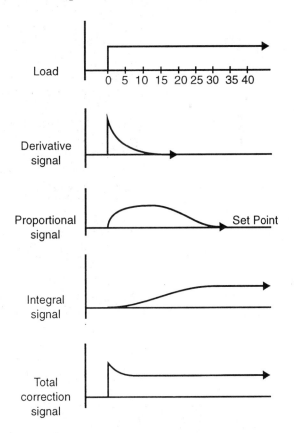

Figure 12-11 Full PID action with long-term load.

zero instantly; but derivative action is not designed to do this. It surges upward in somewhat of an instantaneous manner but dies away quickly yet not instantly. This helps the derivative signal to perform its function in a more useful manner.

3. Note that the integral signal rose very slowly at first, but as time passed, began to rise more rapidly, finally becoming the *total* correction signal and taking care of the long-term load alone.
4. Note that the proportional signal performed as expected at first and then, as a result of the integral signal handling the long-term load, was able to return to the set point. The proportional action is now free to act about its set point to take care of short-term fluctuations in the load (along with the derivative signal, of course).

Proportional Control with Derivative and Integral Action (PID)

The example in Figure 12-11 is proportional control with derivative and integral action, more commonly referred to as *PID*. This is the most common action of all in process control. If one understands PID, then one can proceed to any other facet of the subject of industrial process control. There are a couple of other points the authors would like to clear up, however, before moving on to implementation of the aforementioned control actions.

It should be noted that the process control system *never* affects the load. The process control system controls the measured variable, not the load. The size of the load depends on what the industrial setting demands. The control system is there to *service* the load, not change it. It should also be noted that loads do not necessarily exhibit a sudden jump as shown in all the examples given thus far. This sudden jump was shown simply to keep the examples more understandable.

Finally, the student should be aware that there are many more control modes than P, PI, PD, and PID. These other modes are very often a version of PID with a few more complex actions added. PID, however, is the foundation of almost all control operations; and a fundamental knowledge of how it works serves as a suitable basis for any further study of the subject of process control.

Adjustment of the Proportional Band

In Figure 12-7, the controller has caused the actuator to overcorrect (sometimes referred to as overshoot) in the positive direction. The controller senses this and has the actuator try to correct it; but again the actuator overshoots, this time in the negative direction. In this latter case, the measured variable has been driven below the set point. This process happens once again; and finally, the measured variable is stabilized at the set point. The actions pictured in Figure 12-7 are acceptable and represent reasonable gain or proportional band settings for the controller. (Adjustments to the gain would probably be made, however, to get rid of as much overshoot as possible.) Incorrect proportional band settings for a controller are represented in parts of Figure 12-12.

In Figure 12-12a, the gain of the controller is essentially too small. It cannot open the actuator enough to ever correct for the change in the measured variable. In Figure 12-12b, the gain of the controller is basically correct, as was pictured in Figure 12-7. Figure 12-12c shows a case where the gain of the controller is too great. It continually overdrives the actuator with the result that it takes much longer than necessary or practical to achieve the set point. In fact, if the gain is higher than indicated in Figure 12-12c, the set point will never

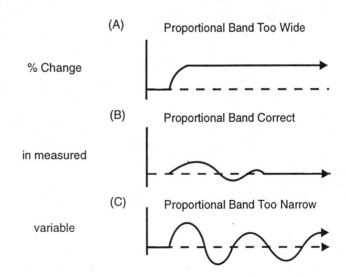

Figure 12-12 Examples of how different proportional band settings can affect the change in the measured variable as the controller attempts to return the measured variable to the set point.

be reached. The actuator will eventually simply be driven continually back and forth between its fully open and fully closed positions. Proper controller gain is important. (Note also that integral reset needs to be present to actually reach the set point.)

Mention was made earlier in this chapter of lag time. A graphical example of lag time is shown in Figure 12-13. The amount of time between point 1, where the measured variable deviated from the set point, and point 2, where the measured variable returns to the set point for a continuous period, is the lag time. Lag time is caused by a number of things. One is the volume of the material whose measured variable is being controlled. For example, at any instant of time, the temperature of a hot-water system will vary from point to point in the containing system if the system is under use. The temperature at the sensor will not necessarily be the temperature at the water inlet or outlet or at many other parts of the water tank or pipe coil. Also, the walls of the pipe coil or tank store energy; and the amount of this energy being absorbed or released is always one of the factors affecting the temperature of the water.

Another property affecting lag time is the time needed to carry a change from one part of the process loop to another. Again, using the hot-water system as an example, it takes time for the temperature sensor to become the same temperature as the surrounding water. Most sensors, especially temperature types, are encapsulated in probes, which protect the sensor from corrosion and breakage and which make it easier to swap sensors whenever needed. It takes time for the material of the probe to come to the same temperature as the surrounding water. Another slow part of the loop is usually the actuator. It takes a finite amount of time for an actuator to move to its newly directed position. This is especially true because most actuators are at least partially mechani-

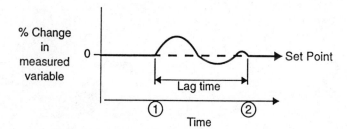

Figure 12-13 Lag time is the time required to return the measured variable to the set point.

cal. Also, pneumatic controllers are inherently slower than electric controllers. All of the factors in the two preceding paragraphs add up to give a lag time.

IMPLEMENTATION OF CONTROLLER ACTION

This portion of the chapter discusses how the three most common implementations of process control—analog electronic, pneumatic, and digital electronic—are carried out.

Analog Electronic Control

On/Off Action. On/off action is most typically performed by an on/off sensor/switch, such as a thermostat composed of a spring-type bimetallic temperature sensor with a glass bulb containing mercury mounted on the top. The wires from the mercury bulb are connected to an electrical relay. The on/off sensor/switch will output only a small amount of voltage or current but enough to switch a relay on or off so that the relay can allow however much electrical power is needed to be supplied to the electrical motors that operate the process. If the process requires fluid instead of electricity, then the relay would perhaps activate a solenoid valve. An electrical power relay is shown in Figure 12-14, and a typical solenoid valve in Figure 12-15.

PID Action. A loop diagram showing proportional control with integral and derivative action is shown in Figure 12-16. The measured variable signal less the set point is often called the error signal. The measured variable signal is fed to the *com-*

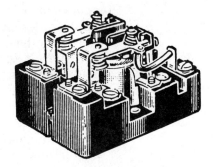

Figure 12-14 A typical electrical relay *(From Rex Miller and Fred W. Culpepper, Jr.,* Electricity and Electronics, *2nd ed., © 1991, Delmar Publishers, Inc., Albany, N.Y.)*

Figure 12-15 An electrical solenoid-actuated valve for fluid control *(Courtesy of D. Gould Co. Inc.; from Jack W. Chaplin,* Instrumentation and Automation for Manufacturing, *© 1992, Delmar Publishers, Inc., Albany, N.Y.)*

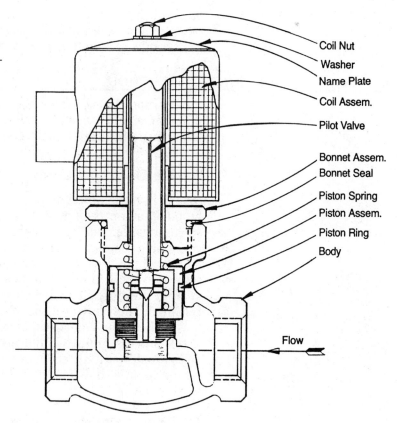

parator, a device that compares two signals and outputs the difference. (Since the comparator effectively does a subtraction process, it is designated with a "–.") The comparator in this case compares the measured variable signal to the set point and feeds the resulting error signal to all three primary components of the controller—the amplifier, the integrator, and the differentiator. The resulting signal from each unit is then fed to the *adder*. The adder is the block with the "+" in the center. It adds the three electrical signals from each of the three primary components. The resulting correction signal is sent to the actuator.

The three primary components are typically built into one unit. Figure 12-17 shows a simple op-amp proportional amplifier with another op-amp at the front to serve as the *comparator.* As mentioned earlier, the comparator compares the incoming measured variable signal to the constant set point

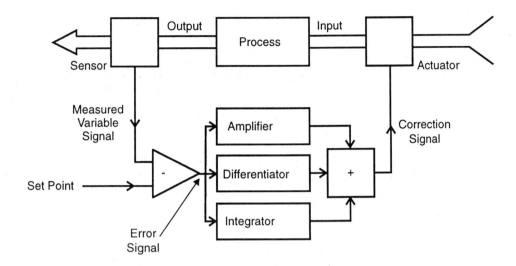

Figure 12-16 Block diagram of analog electronic process control system with PID action.

Figure 12-17 Electronic schematic of proportional control action.

and outputs the difference between the two. This is easy for an op-amp to do, since it has two inputs already designed to perform such a chore. (See the section on op-amps in Chapter 4.)

The difference between the two signals, the error signal, is sent to the op-amp amplifier. The output is sent to the adder—in the case of PID, PI, or PD—or directly to the actuator—in the case of proportional-only action.

Differential action can be achieved with the same op-amp amplifier shown in Figure 12-17 by adding a capacitor into the input of the amplifier (after the comparator). By adjusting the value of the capacitor, the time constant of differentiation can be varied to account for whatever load time is to be considered as very quick load change.

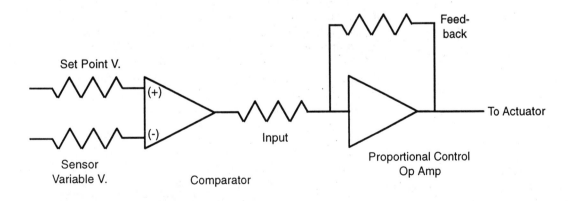

Likewise, differential action can be achieved with the same op-amp configuration as shown in Figure 12-17 by adding a capacitor and parallel resistor into the feedback loop. A total three-mode PID controller is shown in Figure 12-18, except a comparator would need to be added at the front end.

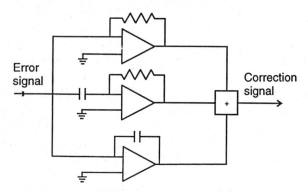

Figure 12-18 Electronic schematic of PID control action (comparator not shown).

Pneumatic Control

Proportional Control. The most common way of carrying out proportional control with pneumatic signals is the use of four bellows that "fight" against one another, either by the use of the beam-and-fulcrum method or by the use of a single round plate directly above all four bellows. The beam-and-fulcrum method is pictured in Figure 12-19.

The items labeled *1, 2, 3,* and *4* are metal bellows. As the air pressure in a bellows increases, the bellows expands. As a bellows expands, it tends to rotate the beam about the fulcrum. Since it is the pressure in the bellows that actually causes movement of the beam, the notation for bellows #1 shall be P1; for bellows #2, P2; etc. Notice that P2 and P4 both tend to move the beam clockwise about the fulcrum, while P1 and P3 tend to move it counterclockwise. The fulcrum is adjustable along the horizonal axis; hence, P2 and P3 are both a distance x away from the fulcrum, and P1 and P4 are a distance y. All bellows have the same cross-sectional area; hence, it can be shown (and will be left as an exercise for the student) that, when in torque equilibrium about the fulcrum:

$$P1 = \frac{x}{y}(P2 - P3) + P4 \qquad (12\text{-}1)$$

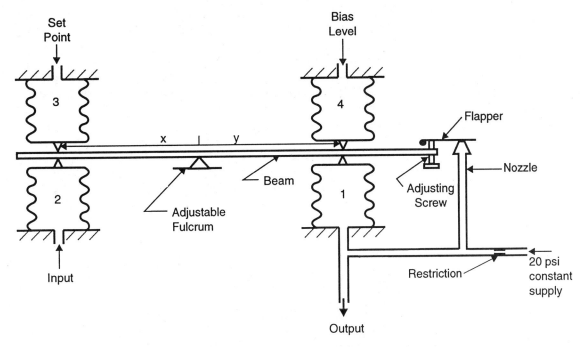

Figure 12-19 Beam-and-fulcrum pneumatic proportional control system.

Notice that if the bellows are connected as indicated in Figure 12-19, then Equation 12-1 makes very good sense and, in regard to proportional amplification, is relatively easy to analyze. In Equation 12-1, P3 is subtracted from P2. From Figure 12-19, it can be seen that P2 is the input signal and that P3 is the set point signal. Thus, P2 – P3 represents the error signal. By looking at Equation 12-1, it can now be seen that the output, P1, is proportional to the amount of error, just as it should be. The factor x/y represents the gain of the amplifier. In other words, the error is multiplied by the gain x/y. From the representation of x and y in the figure, it is obvious that the gain can be controlled by moving the fulcrum left or right.

Finally, P4 represents what is known as a *bias level*. When a signal is made to fluctuate within a given range but the center of that range is not zero, then the signal has a constant bias that is being added to it. The bias level is normally the difference between the center of an allowable signal range and zero. For example, the bias level for the common pneumatic signal range of 3 to 15 psi is 9 psi. (The bias level is somewhat related to the zero, span, and offset of a transducer.)

The remaining question about Figure 12-19 is this: How does one get the system into equilibrium at the proper output signal range of 3 to 15 psi? This is done by adjusting the fulcrum position and the flapper position. Notice that there is a constant pressure of 20 psi being supplied through a restriction to the nozzle and to P1. No air flows out of the bellows, and basically very little air flows out of the output. (Pneumatic signals, although they vary in pressure, are basically static in regard to air flow.) When the flapper rests on the nozzle, there is practically no air flow through the restriction and hence very little pressure loss across the restriction. The pressure in P1, which is equal to the pressure at the output, will be close to the supply pressure, say, in the range 19 to 15 psi.

On the other hand, when the flapper is totally away from the nozzle, there is considerable air flow through the nozzle and considerable pressure loss across the restriction. The resulting pressure P1 will be about 0 to 5 psi. The system is set to equilibrium by setting the set point to the middle of the expected input range (the set point is most commonly set in the middle of the input signal range) and the bias level to 9 psi (the center of the sought-after output range of 3 to 15 psi). The lowest input signal expected is sent to P2 and the fulcrum (or flapper) adjusted to output 3 psi. Then, the highest expected input signal is connected to P2; and the flapper (or fulcrum) is adjusted for an output of 15 psi. This latter adjustment will slightly upset the first adjustment, and back-and-forth adjustments will have to be made. Finally, the following equation will hold:

$$P_{out} = G(P_{in} - \text{set point}) + 9 \qquad (12\text{-}2)$$

where

$$P_{out} = P1$$

$$P_{in} = P2$$

$$G = x/y$$

One of the key points to understanding Figure 12-19 is to understand how the nozzle-and-flapper system works. Notice that a constant 20 psi air supply is provided to the nozzle and bellows #1.

CHAPTER 12 / **297**

PID Pneumatic Control. PID pneumatic control makes use of many more pneumatic components than indicated in the simple proportional control model discussed earlier. These include pneumatic relays (which are oftentimes present in the output section of proportional control–only controllers and increase the accuracy of the control), nested bellows (commonly known as accumulators), and restrictive leak valves.

An example of the flapper/nozzle output system of a beam-and-fulcrum controller is shown in Figure 12-20. The beam

Figure 12-20 Pneumatic PID control system *(From Jack W. Chaplin,* Instrumentation and Automation for Manufacturing, *© 1992, Delmar Publishers, Inc., Albany, N.Y.)*

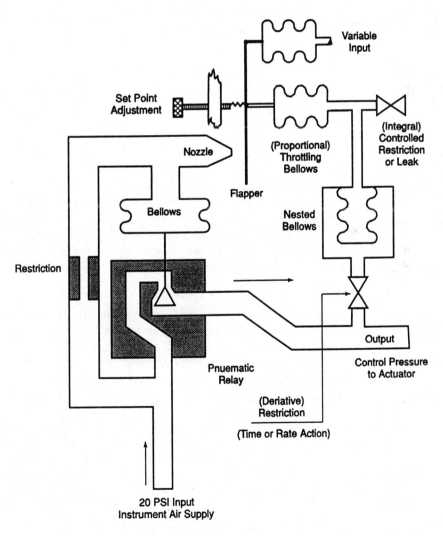

and fulcrum and the bellows P2, P3, and P4 are represented by the one bellows labeled "variable input." The nested bellows (accumulator) with the restrictive valve allows slow accumulation over time of the error signal. The other restricted valve allows quick, direct flow to the output directly from the opposite side of the bellows in the accumulator, depending on the setting of the restricted valve. This temporarily reduces the pressure in the integral nested bellows (inside the accumulator); but the integral action and the derivative action perform exactly opposite tasks, so this makes sense. Also, the integral bellows will slowly develop pressure as long as there is an output signal above the set point.

The two preceding paragraphs do not give, and are not intended to give, an in-depth analysis of Figure 12-20. Detailed analysis of Figure 12-20 is beyond the intended scope of this textbook. The few comments made are intended to give the student at least some idea of the pneumatic PID system so that further study can be more easily accomplished if needed.

Digital Electronic Control

Digital electronic control involves the use of "smart" integrated circuits known as *processors*. There might be a mainframe computer providing the controller function, or there might be a microcomputer or even a more simple programmable controller. There are two very common types of industrial digital controlling, however, and those two methods are discussed here.

Programmable Logic Controller (PLC). Programmable logic controllers (Figure 12-21) are basically digital computers that use *ladder logic*, a programmable logic that is easily understood by engineers and other technical people. They were designed for industrial use and are used to control many automated processes in industry other than process control. They are considered to be relatively simple units insofar as digital computers are concerned and are typically dedicated units; that is, if they perform as a controller for a process control system or for a robot, they have no other functions.

Besides being relatively easy to program, other advantages of PLCs are that they handle pulsed or digital I/O very easily, they are less expensive on the average, and they endure industrial surroundings well. Also, because of their digital I/O

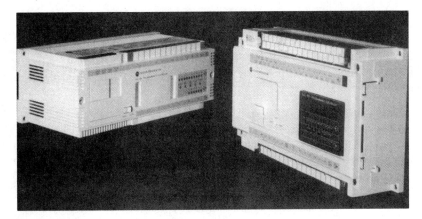

Figure 12-21 Two different models of programmable logic controllers *(Courtesy of Allen-Bradley Company; from Jack W. Chaplin,* Instrumentation and Automation for Manufacturing, *© 1992, Delmar Publishers, Inc., Albany, N.Y.)*

advantage, they are generally regarded as being very fast. Their biggest disadvantage is that they are designed for stand-alone applications and thus do not work as well as other types of systems for central process control.

Distributed Control Systems (DCS). Distributed control systems usually make use of computers—usually on the order of a microcomputer—at each station of a multiloop multicontrol system. Usually, these individual station computers are more reliable in regard to accurate communications than a PLC. (Also, it should be noted that PLCs can be used in place of the microcomputers and that even the noncomputer types such as pneumatic and analog electronic controllers can be monitored by a central system.) Each small computer is under the control of one larger central computer. Central control of many of the automated operations in a manufacturing plant is oftentimes preferred, and DCSs provide this.

Other advantages of DCSs is that the individual station computers can handle more complex programs than a PLC and thus can handle more complex control actions than simply PIDs. Also, the individual computers enable the central computer to display superb graphics (see Figure 12-22) representing various control loops that the DCS system is operating.

The next move in the future is probably going to be more use of PLCs in DCS systems. PLCs are continually being improved, they have some advantages mentioned earlier, and it is currently thought that one of the next systems of the future

Figure 12-22 Central control computer for a DCS system *(Courtesy of Leeds & Northrup Co., A Unit of General Signal; from Jack W. Chaplin,* Instrumentation and Automation for Manufacturing, *© 1992, Delmar Publishers, Inc., Albany, N.Y.)*

is going to be DCS with individual stations of improved PLCs (when needed) and microcomputers otherwise. Although most of the direction of process control research is currently toward improved centralized plant control, there will always be a place for stand-alone pneumatic controllers and analog electronic controllers in the industrial workplace.

REVIEW MATERIALS

Important Terms

controller action
differential on/off
two-position control
process load
derivative action
slope
anticipatory action
reset action
correction signal
adder
bias level
flapper
DCS

on/off control action
lag time
proportional action
proportional band
rate action
integral action
long-term load
error signal
PID action
comparator
nozzle
PLC
overshoot

Questions

1. Based on the results to date of the heating or cooling of the room in which this course on basic industrial instrumentation is being given, what type of controller action is being used?
2. In regard to a home central electrical heating system, what portion of the system would be considered to be the controller? The actuator? The sensor?
3. Which system would be more practical for controlling the level of coal in a silo: on/off control or differential on/off control? Discuss.
4. What is the difference between amplification factor and gain? Discuss.
5. In regard to a heat-transfer system where steam is used to heat water, would the amount of steam needed be considered to be the *load*, or would it be the amount of hot water needed? Discuss.
6. Describe a home central air conditioning system that has proportional action. What is one of the ways it could be designed? Why is proportional action not normally used for home air conditioning?
7. Would a proportional band that was set by a novice technician to have a percentage error range of +200% to −200% with the same percentage output range as shown in Figure 12-4 have more of a tendency to overshoot the set point and even perhaps go into uncontrolled oscillation, or to never be able to reach the set point even a single time? Discuss.
8. What gain would the proportional band in Question 7 represent?
9. What is the difference between an error signal and a measured variable signal?
10. What is the difference between a correction signal and a manipulated variable signal?
11. What are some of the factors that keep a process control system from performing instantaneous corrections? What are some of the actions that could be taken to attempt to reduce the correction time?
12. Why would derivative action sometimes be referred to as rate action? As anticipatory action?
13. Why would integral action sometimes be called reset action?
14. Describe a process control situation where 24 hours might be the minimum definition of "long term."
15. Very-short-term load fluctuations would be best taken care of by proportional action, integral action, derivative action, or by what combination of these?
16. If the fulcrum in Figure 12-19 was moved to the left, would the gain increase or decrease? Would the bias level increase or decrease?
17. Discuss why the restriction is necessary in Figure 12-19.
18. Although ladder logic is not a programming language that is commonly known, why is it preferred for programmable logic controllers?
19. What is the main function of the individual station computers in distributed control systems?

Problems

1. Prepare a graphical diagram similar to Figure 12-2 that represents a typical home central air conditioning system using differential on/off control action.
2. Draw a graphical diagram similar to Figure 12-6 that represents a "perfect" proportional control system—one that has a perfectly set proportional band with a gain of 1.5 and that has no lag time. Use the load in Figure 12-23.

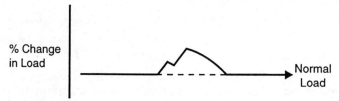

Figure 12-23 Example of a short-term fluctuation in load.

3. Draw a graphical diagram similar to Figure 12-7 that represents an ordinary proportional control system—one that has a small amount of overshoot from a proportional band with a gain of 2 and a typical lag time as in Figure 12-6. Use Figure 12-24 for the load.

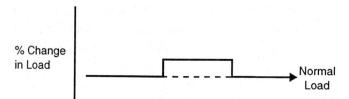

Figure 12-24 Example of a fluctuation in load. The time could be short term or long term, depending on the process (for use in various problems).

4. Repeat Problem 3 for a proportional band that is extremely narrow.
5. Plot the slope of the measured variable for Figure 12-25a and b.
6. Draw a complete graphical PD diagram similar to Figure 12-11 for the load shown in Figure 12-24.
7. Plot the integral signal (the sum of the area under the measured variable curve) for both parts of Figure 12-25.
8. Draw a complete graphical PID diagram similar to Figure 12-13 for the load pictured in Figure 12-24.
    a. Assume that long-term load was never even close to being reached.
    b. Assume that long-term load was reached just as the load went back to normal.
    c. Assume that long-term load was reached at midway through the load shown in Figure 12-24.

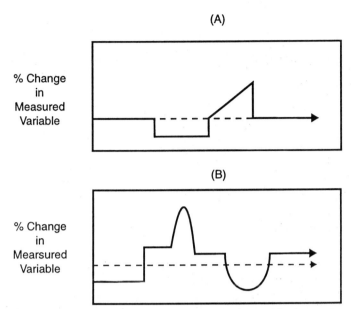

Figure 12-25 Percentage changes in measured variable (for use in various problems).

9. Since the output of the proportional amplifier in an analog electronic PID controller is proportional to the error signal, the integrator and the differentiator could both take their input signals from the output of the amplifier rather than from the output of the comparator. Draw a block diagram similar to Figure 12-16 showing such a set-up.

10. Derive Equation 12-1 for a beam-and-fulcrum system as shown in Figure 12-19 in an equilibrium (static, no rotation, all torques summed to zero) condition. Assume each bellows has an equal cross-sectional area and that each has an internal pressure of P1, P2, P3, and P4, in turn.

11. From Equation 12-2, calculate P_{out} if x = 4 inches, y = 6 inches, set point = 3 psi, and where 5 psi is connected to bellows P2.

12. If one wished to use the pneumatic amplifier in Figure 12-19 to convert the input signal range of 0 to 6 psi to the standard output range of 3 to 15 psi, what gain and what set point would be needed?

Instrument Society of America (ISA) Symbols

Following are examples of symbols used in piping and instrumentation diagrams (PID). A complete set of diagrams is available from ISA.*

GENERAL INSTRUMENT SYMBOLS

Figure A-1 Field-mounted discrete instrument.

Figure A-2 Field-mounted computer function.

Figure A-3 Instruments sharing common housing.

CONTROL-VALVE BODY SYMBOLS

Figure A-4 Butterfly.

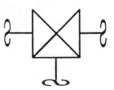

Figure A-5 Three-way.

*Reprinted by permission. Copyright © 1984, Instrument Society of America. From ANSI/ISA-S5.1-1984—*Instrumentation Symbols and Identification.*

ACTUATOR SYMBOLS

Figure A-6 Diaphragm, pressure-balanced.

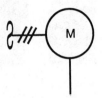

Figure A-7 Rotary motor (shown typically with electric signal; may be hydraulic or pneumatic).

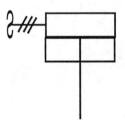

Figure A-8 Spring-opposed single-acting.

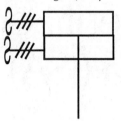

Figure A-9 Double-acting.

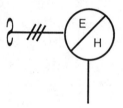

Figure A-10 Electrohydraulic.

Figure A-11 Hand actuator or hand wheel.

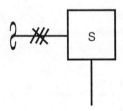

Figure A-12 Solenoid.

Figure A-13 For pressure relief or safety valves only; denotes a spring, weight, or integral pilot.

SELF-ACTUATED DEVICE SYMBOLS

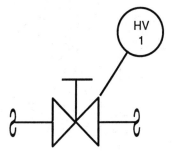

Figure A-14 Hand control valve in process line.

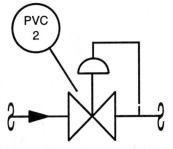

Figure A-15 Pressure-reducing regulator with external pressure tap.

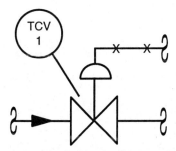

Figure A-16 Temperature regulator, filled-system type.

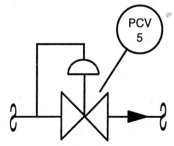

Figure A-17 Backpressure regulator with external pressure tap.

SYMBOLS FOR ACTUATOR ACTION IF POWER FAILS

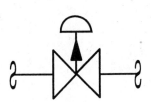

Figure A-18 Two-way valve, fail open.

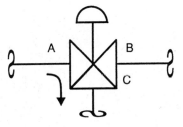

Figure A-19 Three-way valve, fail open to path A–C.

PRIMARY ELEMENT SYMBOLS

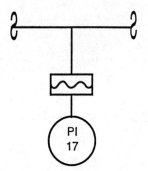

Figure A-20 With pressure lead line.

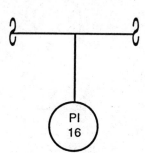

Figure A-21 Pressure indicator, direct-connected.

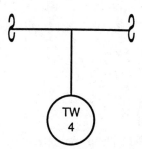

Figure A-22 Temperature connection with well.

MISCELLANEOUS SYMBOLS

Figure A-23 Pilot light.

Figure A-24 Reset for latch-type actuator.

APPENDIX B

Answers to Odd-Numbered Problems

CHAPTER 1

1. Process: Maintaining a constant fluid level. Measured variable: The fluid level. Manipulated variable: Flow of fluid out. Controlled variable: Same as the measured variable by definition. Controller: The operator (or perhaps, the operator's brain). Actuator: The valve. Sensor: Eyesight combined with sight tube.
3. Say the thermocouple output is 100 μv. A = 100,000
5. ±4% FS
7. ±1% FS, ±0.88% Span
9. ±0.01v would be the best guess without having any further knowledge. ±0.005 v.
11. Hysteresis seems to be present. The single data point at 10.0 true-pressure-decreasing can be ignored because of substantial evidence that hysteresis is present.

CHAPTER 2

1. a. 470 ± 10% ohms
   b. 33 ± 20% ohms
   c. 92,000 ± 5% ohms
3. a. 0.1 amp
   b. 1.67 amps

5. a. 1.20 amps
   b. 0.47 amp
   c. 2.35 volts
   d. 5.64 volts
   e. 4.03 volts
7. 3.43 volts, 13.71 volts, 20.57 volts
9. a. 6 volts
   b. 8 volts
   c. 2.4 volts
11. Decrease to 45 ohms. 0 volts. −0.32 volt
13. 1 Megohm
15. $I_1 = 9$ amps, $I_2 = -1.5$ amps, $I_3 = 10.5$ amps

CHAPTER 3

1. 1.77×10^9 ohms, 1.77×10^6 ohms
3. 1.77 mA
5. 3.21 amps
7.

CHAPTER 4

1. See Figure B-1.
   No, the op-amp will go into uncontrolled oscillation.
3. One method would be to place a high-resistance pot between the +15 v and -15 v power leads and then connect the wiper to the input through a resistor.
5. See Figure B-2.
7. 75,000
9. The convertor has no gain, since input units and output units are different. 45,000 v/A.

CHAPTER 5

1. 10.83 psi
3. 86.7 psi
5. 193.1 kPa, 414 kPa, 89.6 kPa
7. 138.5 inches, 16.62 inches

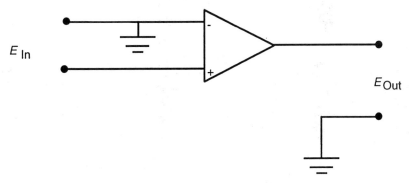

Figure B-1 Answer to the first part of Problem 4-1.

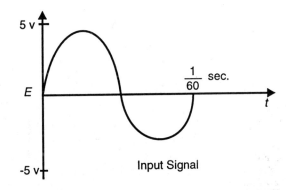

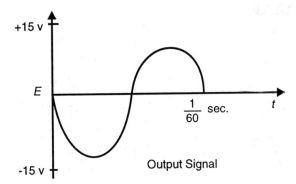

Figure B-2 Answer to Problem 4-5.

9. 3.00 psi, 216 psf
11. 34.2 psia
13. 124.8 lbs/ft^3

15. 11.05 psia
17. 48,400 lbs
19. 148.0 psia
21. 500 psi
23. 11.81 psi
25. 624 pounds on every square foot
27. 7.07 psi, 20,000 lbs
29. 1.603 ft³, 149.7 lbs/ft³

CHAPTER 6

1. 9.97 v, 9.85 v
3. 17,857
5. One set would be $R_1 = R_2 = 100\ \Omega$, $R_3 = 500\ \Omega$
7. 250 Ω
9. See Figure B-3.

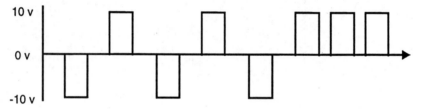

Figure B-3 Answer to problem 6-9.

11. See Figure B-4.

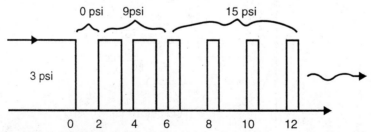

Figure B-4 Answer to problem 6-11. If E is halved, the amplitude of the pulses will be halved.

13. $k = \dfrac{f_{out}}{E_{in}} = \dfrac{600\ \text{Hz}}{10\ \text{v}} = 60\ \text{Hz/v}$

CHAPTER 7

1. 22.2°, −35.6°, 150°
3. 540°, 480°, 450°
5. 253°, 293°, 393°
7. 55°
9. 16.67 Celsius degrees
11. 303°
13. 476°
15. 767 BTUs/hr
17. 48.3° F
19. 0.0004 inch per inch
21. 76.55 ft
23. 0.00000521/° F
25. 200° to 900° F
27. About 10 seconds
29. 186.5° F

CHAPTER 8

1. 23.1 ft
3. 0 to 5 psi gage
5. 51.1 lbs
7. 1.097
9. 13.00 psi
11. 1.950 psi
13. 19.60 lbs
15. 4.54 to 22.7 lbs

CHAPTER 9

1. 1.419 fps, 5.67 fps
3. 81.8 N/sec
5. 1.022 fps
7. 0.05 m^3/sec
9. 61.1 fps, 137.6 fps
11. 16.05 fps
13. 139.0 fps

15. 80.8 psi
17. 21.8 psi
19. 3.66 fps
21. 0.0774 ft
23. 80
25. 10 fps
27. 7.78 gpm
29. 0.959 fps

CHAPTER 10

1. 78%, 56%, 21%
3. 67.0° F, 75° F
5. 76, 125, and 110 grains/lb
7. 39%, 134 grains/lb
9. 0.14 psi, 0.28 psi
11. 20.5 BTUs/lb
13. 77 grains/lb
15. 28.2 BTUs/lb, 158 grains/lb

CHAPTER 11

1. 0.984 slug/ft^3
3. 1.938 slugs/ft^3
5. 5.24
7. 0.01111 lb sec/ft^2
9. 0.000333 lb sec/ft^2
11. 0.0100 lb/ft^2
13. 64.4 lbs
15. 6.67, 12
17. 1.781 R
19. 9.54
21. 0.00632 N/m^2
23. 103.0
25. 0.001 g/l, 0.1 g/l
27. 1.995 × 10^{-11} g/l
29. 1,000

APPENDIX B / 315

CHAPTER 12

1. See Figure B-5.

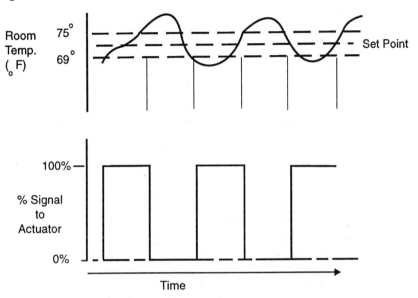

Figure B-5 Answer to problem 12-1.

3. See Figure B-6.

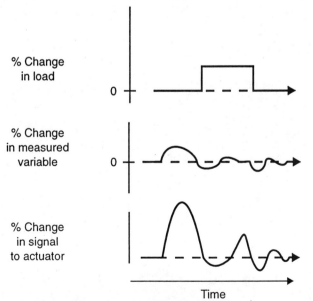

Figure B-6 Answer to problem 12-3.

5. See Figure B-7.

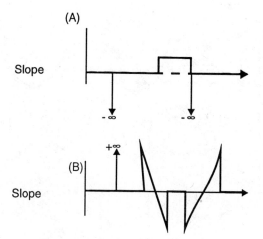

Figure B-7 Answer to problem 12-5.

7. See Figure B-8.

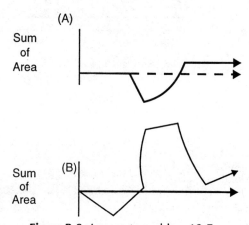

Figure B-8 Answer to problem 12-7.

9. See Figure B-9.
11. 10.3 psi

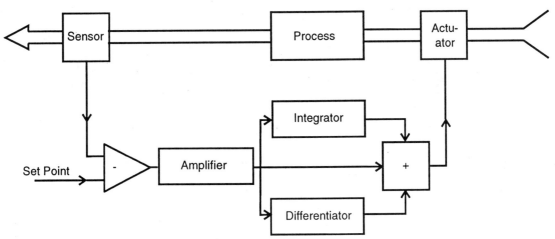

Figure B-9 Answer to problem 12-9.

Glossary

absolute accuracy accuracy stated as a definite amount (i.e., not a percentage)
absolute position measurement position measured from a fixed point
absolute pressure pressure measured from a perfect vacuum; a true "zero" pressure
accuracy measure of the difference between the indicated values from an instrument and the true values
active device an electrical device that amplifies or controls
actuator device that performs an action on one of the input variables of a process according to a signal received from the controller
adder a device that sums two simultaneous inputs and then outputs this sum
alternating current current that flows one direction during half of a regular time period and the opposite direction during the other half
ammeter a device used to measure current
ampere coulombs per second; the unit of current
amplifier an electrical circuit that modifies its input so that more power, usually in the form of an amplified input signal, is available at the output
analog continuous
aneroid barometer barometer which uses a capsule from which all the air has been removed as a sensing element
anticipatory action see **derivative action**
aqueous solution solution containing water
atmospheric pressure pressure acting on objects on the earth's surface caused by the weight of the air in the earth's atmosphere

barometer instrument used for measuring atmospheric pressure
battery two or more cells connected in series
Bernoulli equation flow equation which includes velocity, pressure, and elevation terms

beta ratio   ratio of restriction diameter to diameter of pipe
bimetallic   thermometer with sensing element made of two metals with differing expansion rates
binary   two values; a number system using base 2
bit   a binary digit
boundary layer   the layer of fluid nearest the boundary in fluid flow
Bourdon tube   metallic tube often used as sensing element in a pressure gage
bridge   a set of resistances in parallel arranged so that the output is taken from a meter that, when indicated on a schematic diagram, bridges or spans the network; a circuit designed so that small changes in resistance are easily detected
British Thermal Unit   measure of heat; the amount of heat required to raise one pound of water one Fahrenheit degree at 68° F and atmospheric pressure
buffer amplifier   a circuit for matching the output impedance of one circuit to the input impedance of another circuit
buoyancy   the upward force on an object floating or immersed in a fluid; caused by the difference in pressure above and below the object
byte   eight bits

calorie   measure of heat; the amount of heat required to raise the temperature of one gram of water one Celsius degree
capacitance   measure of a device's ability to store electrical charge per amount of voltage applied to the device
capacitance probe   instrument using the capacitance between two objects to measure level
capacitor   a device that can store electrical charge if a voltage is applied across it
cell   a simple power source that provides voltage, usually by means of a chemical reaction
Celsius   one of the commonly used temperature scales
coefficient of heat transfer   term used in calculation of heat transfer by convection
coefficient of thermal expansion   term used to determine the amount of expansion due to heating or cooling
comparator   device which compares two signals and outputs the difference
concentric plate   orifice plate with hole located at its center
conduction   the movement of heat from one molecule of an object to another
conductivity probe   instrument using two electrodes to measure level
conductor   a material with electrons that are free to move under the influence of voltage
continuity equation   flow equation which states that, if the overall flow rate is not changing with time, the flow rate past any section of the system must be constant
continuous level measurement   gives level location at any instant in time
controlled variable   variable measured to indicate condition of the process output
controller   process control loop element that determines what action is to be taken
convection   the movement of heat by the motion of warm or hot material
converter   device that changes signal representation mode but not the type of energy used as the signal carrier
correction signal   manipulated variable signal
current   the rate of electron flow

DC amplifier an amplifier that modifies DC as well as AC signals
DCS distributed control system
dead weight tester device for calibrating pressure-measuring devices which uses weights to provide the forces
decibel (dB) a unit used to compare sound levels
density the amount of mass in a unit volume
derivative action action proportional to the rate with which the measured variable is changing
dew point temperature at which condensation begins to take place
dielectric constant factor by which the capacitance between two plates changes when a particular material fills the space between the plates
differential amplifier an amplifier that amplifies the difference between two simultaneous inputs
differential on/off control process actuator does not go "on" or "off" until the measured variable reaches a small percentage above or below the set point
digital discrete
direct level-measuring device works directly with the fluid level
dry-bulb temperature temperature indicated by a thermometer whose sensing element is dry
dynamic pressure that part of the total pressure in a moving fluid caused by the fluid motion
dynamometer instrument used for measuring torque or power

eccentric plate orifice plate with hole located below its center to allow for passage of suspended solids
effective value an amount of DC voltage or DC current that would effectively produce the same amount of real power as the actual AC voltage or AC current being measured
electrical noise undesired electrical voltages
electromagnetic flow meter flow-measuring device which senses the change in a magnetic field between two electrodes as flow passes between them
electromagnetism the interrelationship between magnetism and electric current
electromotive force voltage; a force that causes electrons to move
electronics the study of active devices and the circuits that incorporate them
elements devices which are part of the process control loop, such as sensors, transducers, actuators, controllers

error signal difference between measured signal value and set point
Fahrenheit one of the commonly used temperature scales
farad a unit of capacitance
feedback (1) voltage fed from the output of an amplifier to the input in order to control the characteristics of the amplifier; (2) the measured variable signal fed to the controller in a closed-loop system so as to closely control the output of the controller
fiber optics transmission of information through the use of light signals traveling through special optical paths
flow nozzle device placed in flow line which provides a pressure drop that can be related to flow rate

flow rate   amount of fluid passing a given point in any given interval of time
flume   open-channel flow-measuring device
form drag   force acting on an object from fluid impacting on the object
free convection   movement of heat as a result of density differences
free surface   the surface of the liquid in open-channel flow that is in contact with the atmosphere
frequency   number of cycles completed in one second; alternatively, 2π times the number of cycles completed in one second (latter definition is sometimes referred to as *rotational frequency*)

gage pressure   pressure indicated by a gage; the reading is the amount of pressure at the gage location above atmospheric pressure
gain   the ratio of output to input when units of both are the same; amplification factor
ground   the voltage level of the earth, used as a common reference point for many circuits

head   sometimes used to indicate pressure; one foot of "head" for water is the pressure under a column of water one foot high
heat   a form of energy; related to motion of atoms or molecules
heat transfer   the study of heat movement
henry (H)   the unit of inductance
hertz (Hz)   cycles/second
hot-wire anemometry   fluid velocity-measuring device which senses temperature changes resulting from fluid flow
humidity   general term relating to the amount of water vapor present
humidity ratio   mass of water vapor in a mixture divided by the mass of dry air or gas in the mixture
hydrometer   instrument for measuring liquid density
hydrostatic paradox   the fact that pressure varies with depth in a nonmoving fluid and does not vary at any given depth
hydrostatic pressure   pressure caused by the weight of a nonmoving fluid
hygrometer   relative humidity-measuring device
hygroscopic   material whose conductivity changes with moisture content
hysteresis   nonreproducibility dependent on direction with which the measurement approaches the true value

impact pressure   the sum of the static and dynamic pressure in a moving fluid
impedance   an opposition to AC or DC current flow caused by resistance, inductance, or capacitance
incremental position measurement   position measurement in which correct reading is not retained if power is lost
indirect level-measuring device   works with something other than the level itself
inductance a measure of the tendency of a device to form a magnetic field when carrying a current

inductor a device that forms a relatively large magnetic field when carrying a current
instrument a device used to measure a physical variable
insulator a material whose electrons are so tightly bound that they cannot move under the influence of voltage
integral action action designed to correct for long-term loads
integrated circuit a circuit formed with one unit of a semiconductor crystal

Kelvin the absolute temperature scale associated with the Celsius scale
Kirchhoff's Voltage Law the algebraic sum of voltages around a closed path is zero

ladder logic programmable logic used to control automated industrial processes
lag time time required for control system to return a measured variable to its set point
laminar flow smooth flow in which the fluid tends to move in layers
linear expansion expansion in one direction, such as the length of a wire or rod
linearity measure of the direct proportionality between actual value of the variable being measured and the value of the output of the instrument
load demand placed on process control system for a needed amount of controlled output
load cell device for measuring force or movement
loudness a subjective quantity used to measure relative sound strength

manipulated variable variable changed by the actuator
measured variable variable measured to indicate condition of the process output
meniscus the convex or concave surface of a column of liquid
minor losses pressure losses in a flow system associated with fittings, such as elbows, valves, and the like
moment the effect of a force acting at a given perpendicular distance from a point

natural convection movement of heat as a result of density differences
Newtonian fluid fluid in which the velocity varies linearly across the flow section between parallel plates
node a junction of three or more conductors
noise term usually used to indicate unwanted or undesirable sound
nutating disk meter flow-measuring device using a disk that rotates and wobbles in response to the flow

off state low pulse (zero)
offset the nonzero output of a circuit when the input is zero; the low end of a range
ohm the unit of resistance; one volt per ampere
ohmmeter a device used to measure resistance
on/off control process actuator has only two positions: on and off
on state high pulse (one)
open a break in a circuit
open-channel flow flow in an open conduit (e.g., as in a ditch)

operational amplifier a very common integrated circuit often used as the key ingredient of amplifier circuits and also of many other types of circuits

orifice plate plate containing a hole which when placed in a pipe causes a pressure drop which can be related to flow rate

over pressure term used to describe the amount of pressure a gage can withstand

overshoot overcorrection by actuator

parabolic velocity distribution occurs in laminar flow when the velocity across the cross-section takes on the shape of a parabola

parallel transmission simultaneous transmission from analog to digital

Pascal pressure reading in SI units; newtons per square meter

Pascal's Law pressure applied to an enclosed fluid is transmitted to every part of the fluid

passive device an electrical device that is not an active device

percent reading accuracy given in terms of the reading being made

percentage full-scale accuracy accuracy determined by dividing the accuracy of an instrument by its full-scale output

period a fixed amount of time during which alternating current is completing one full cycle of amplitude and direction; the time is set by the alternating current source such as a generator; the inverse of the frequency in Hertz units

pH a term used to indicate the activity of the hydrogen ions in a solution; helps describe the acidity or alkalinity of the solution

phase term used to describe the state of matter (i.e., solid, liquid, or gas); the fractional part of a cycle through which a periodic variable has advanced at a given instant

phons a unit for describing the difference in loudness levels

PID proportional control with derivative and integral action

piezoelectric effect electrical voltage across points in certain materials when pressure is applied to a crystal of the material

Pitot-static tube device used to measure flow rate using the difference between dynamic and static pressures

PLC programmable logic controllers

poise viscosity measurement unit

position level measurement indicates when a particular level has been reached

pot an abbreviation for potentiometer

potentiometer variable resistance device that uses a wiper to select a voltage

precision the fineness with which an instrument can be read

pressure the strength of a force divided by the area over which it acts; often measured in pounds per square inch

pressure differential the difference in pressure between two locations

process sequence of operations carried out to achieve a desired end result

process control automatic holding of certain process variables within given limits

processors integrated circuits used in digital electronic control

proof pressure the pressure that a pressure-measuring device can tolerate

proportional action percentage change of the manipulated variable signal sent by the controller is proportional to the percentage change of measured variable signal
psychrometric chart chart containing various items dealing with moisture-containing air
pure tone sound of a single frequency
pyrometer instrument for measuring temperature made of a lens and a sensing device

radiation emission of energy from a body in the form of electromagnetic waves
random noise sound coming from more that one source
range lowest and highest readings a sensing device can make
Rankine the absolute temperature scale associated with the Fahrenheit scale
rate action see **derivative action**
reactance an opposition to AC current flow caused by a capacitor or an inductor
reflectance property of a material related to its ability to absorb and scatter infrared waves
regular amplifier an amplifier with no special features other than providing gain (usually adjustable)
relative humidity amount of water vapor present in a given volume expressed as a percentage of the amount that would be present in the same volume under saturated conditions at the same pressure and temperature
reluctance opposition to carrying magnetic flux; the magnetic equivalence to resistance
repeatability measure of closeness of agreement among several consecutive readings
reproducibility ability of an instrument to produce the same measurement over and over again when the conditions return to the earlier state
reset action see **integral action**
resistance a measure of the difficulty of electron movement in a given material
resistivity a temperature-dependent "constant" that reflects a given material's resistance to electron flow
resistor a material that has electrons that can move, but only with a measurable amount of difficulty, under the influence of voltage
resolution smallest amount of variable being measured that the measuring instrument can resolve and indicate or display
Reynolds number dimensionless number describing whether the flow is laminar or turbulent
rotometer flow rate-measuring device in which a float moves in a variable-area tube

saturated condition maximum amount of water vapor is present for the given pressure and temperature conditions
sealing fluid fluid used in a manometer to separate the fluid whose pressure is being measured from the manometer fluid
secondary sound source object that receives sound and reflects part of it
segmented plate orifice plate with hole located so as to allow suspended solids to pass through
semiconductor a material whose resistance varies from low to high depending on impurities present and polarity of applied voltage

sensor device that senses a fundamental physical variable
serial transmission sequential transmission of digital bits
set point fixed signal fed into controller which has a value equal to the signal that the sensor would send if the measured variable were at its desired value
shunt a bypass
signal conditioning modification of a signal so that it can be transmitted more accurately and more dependably
single-point level measurement indicates when a particular level has been reached
sling psychrometer device for measuring relative humidity
sone a unit for measuring loudness
sound pressure level difference between the air pressure at a point at a given instant in time and the average air pressure at that point
span difference between lowest and highest reading for an instrument or a gage
specific gravity the ratio of the specific weight of a solid or liquid material and the specific weight of water; for a gas, the ratio of the specific weight of the gas and the specific weight of air
specific heat amount of heat required to raise a definite amount of a substance by one degree
specific humidity mass of water vapor in a mixture divided by the mass of dry air or gas in the mixture
specific weight the weight of one cubic foot of a material
standing waves a particular effect of combining sounds from primary and secondary sources
static pressure that part of the total pressure in a moving fluid not caused by the fluid motion
stoke viscosity measurement unit
strain gage a resistive path of metal particles cemented on an insulator and used as a displacement-to-electrical transducer
sublimation passing directly from solid to vapor
superconductor a material with no electrical resistance

telemetry (telemetering) electrical transmission of information over long distances by other-than-digital means
temperature term used to describe the hotness or coldness of an object
thermal conductivity measure of the ability of a material to conduct heat
thermal expansion the expanding of a material as a result of its being heated
thermal time constant time required for a temperature-sensing device to change 63.2% of the difference between two different temperatures
thermistor temperature-measuring instrument which uses a metal oxide that experiences a decrease in electrical resistance with increasing temperature
thermocouple temperature-measuring device which uses changes in electrical currents as the sensing mechanism
thermometer instrument used to measure temperature
thermopile thermocouples connected in series

time constant the amount of time needed for a capacitance, C, to discharge through a resistance, R, until the voltage across the capacitance is 37% of the original voltage; the product, RC, of the aforementioned set-up
torque name given for a moment that tends to cause a twisting action
torr pressure caused by the weight of a column of mercury one mm high
total flow amount of flow past a given point over some length of time
total pressure the sum of the static and dynamic pressures in a moving fluid
transducer device that changes energy from one form to another
transfer function equation that describes the relationship between the input and output
transformer a device that transforms the power of alternating current from its primary coil to its secondary coil
transmission transferring of information signals from one point to another
transmitter device that conditions the signal received from a transducer so that it is strong enough to be sent some distance away; a transducer or a converter that performs signal conditioning
turbine flow meter flow-measuring device utilizing a turbine wheel to sense the flow
turbulent flow "mixed-up" flow in which there are random velocity fluctuations on top of the average flow
two-position action process actuator has only two positions: on and off

U-tube manometer glass tube in shape of the letter U; used to measure pressure or pressure differences
ultrasonic probe instrument using high-frequency sound waves to indicate level

vacuum (pressure) usually used to describe the amount that the pressure at a given point is below atmospheric pressure
velocity in flow, the measure of the speed and the direction of movement
vena contracta the narrowing down of the fluid flow stream as it passes through an obstruction
Venturi tube specially shaped section of pipe that provides a pressure drop which can be related to flow rate
viscometer (viscosimeter) instrument for measuring viscosity
viscosity term describing the resistance of a fluid to flow
volt the unit of electromotive force
voltage an electromotive force that causes electrons to move
voltage drop the difference in voltage between two points
volumetric expansion change in volume
vortex swirling or rotating fluid motion

weir open-channel flow-measuring device
wet-bulb temperature temperature indicated by a thermometer whose sensing element is kept moist
Wheatstone bridge the most common bridge circuit

Index

Page numbers in *italic* indicate figures.

Absolute accuracy, 10
Absolute scales, 148–149
Absolute zero, 148–149
Acidity, 265
Accumulators. *See* Bellows, nested
Accuracy, 8, 9–11, 12, 91, 92, 109, 117, 130, 137, 138, 170, 171, 189, 212, 214, 215–216, 217, 221, 222, 245, 253, 297
  absolute, 10
  defined, 9
  percentage full-scale, 9, 10
Active-band sound analyzer, 263, *264*
Actuator, 3, 4, 5, 6, 7, 20, 130, 133, 138, 139, 274, 275, 276, 277, 279, 280, 283, 285, 287, 289–290, 292
  defined, 5
Adder, 83, 87, 292, 293
Alkalinity, 265
Alternating current (AC). *See* Electricity, AC
Ammeter, 12, 36, 38, 40, 69
Amp (A), 21, 22
Ampere. *See* Amp
Amplification, 134
Amplifier, 263, 292, 293, 295
  AC, 86
  buffer, 83, 89–90
  current, 88
  DC, 85–86

Amplifier *(continued)*
  differential, 83, 85, 86–87
  operational, 80–83, 85, 131, 133, 134, 135, 292–293, 294
  schematic symbol, *82*
  regular, 83–85
Amplitude, 65, 130, 136–137, 138, 140, 141, 240
Analog-to-digital converter (A/D or ADC), 92, 138–139
Anemometry, hot-wire, 218
Anticipatory action. *See* Derivative action
Atmospheres (atm), 97, 98

Baffles, 193
Balance, analytical, 256–257
Barometer, 112–113
  aneroid, 112–113
Battery, 24, 28, 41
  defined, 24
Beam-and-fulcrum controller, 294–296, 297–298
Bellows, 110, *111*, 112, 113
  materials for construction of, 112
  nested, 297–298
Bernoulli equation, 200, 203–206
Beta ratio, 211–212
Bias level, 295
Binary digits (bits), 91–92, 136, 138
Bits. *See* Binary digits

Boundary area, 250
Boundary layer, 198
Bourdon tube, 110–111, *111*, 113, 117, 140, 161
  materials for construction of, 111
Bridge
  resistance, 33–34
  Wheatstone, 33–34, 35, 36, 164
British Thermal Unit (BTU), 148, 149, 235
  defined, 149
Bubbler system, 187, 193, 247
Bulb, 159, 162, 163, 164, 171, 291
  location of, 170
  size of, 164, 168
Buoyancy, 102–103, 179, 182, 183, 190, 191, 244, 245
  defined, 102, 179
Byte, 137, 138

Calibration
  direct secondary, 222
  of flow-measuring devices, 222
  of pH-measuring devices, 266
  of pressure-sensing devices, 119–120
  of sound-measuring devices, 263
  of temperature-measuring devices, 171
  of viscosity-measuring devices, 253
Calorie, 148, 150, 151
  defined, 150

329

Cams, 30, 140–141
Capacitance, 53–61, 62, 64, 68, 69, 114, 179, 193, 236. *See also* Capacitors
  defined, 53, 54
Capacitance meter, 55
Capacitive reactance, 55–56, 63, 68
Capacitors, 53–54, 56, 58–59, 60, 61, 62, 64, 65, 68, 80, 85–86, 89, 179, 293, 294. *See also* Capacitance
  in parallel, 56, 58
  in series, 56, 57, 58
Capillary tube, 252
Capsule, 110, *111*, 112
  materials for construction of, 112
Celsius scale, 148, 153, 154
Charles's law, 163
Chassis connection, 28–29
Chip, 80
Circuit, 21, 31–32, 38, 40, 41, 42, 43, 44, 52, 53, 54, 55, 56, 57, 58, 60, 61, 62, 65, 66, 68, 69, 70, 72, 73, 80, 82, 84, 85, 88, 89–91, 92, 115, 131, 133, 217, 263
  bridge, 35–36, 38, 39
  integrated, 80, 93, 298
Closed-loop system, 254
Coefficient of thermal expansion, 152
Comparator, 291–293, 294
Compensation, 170–171
Condensation, 230, 232
Conduction, 151
  heat, 155–156
Conductivity, thermal (k), 151–152
Conductors, 20, 21, 75, 89, 166, 184, 216
Contact method, 182
Continuity equation, 201–202, 205–206
Controlled variables. *See* Measured variables
Controller, 2–3, 5–7, 14, 139, 141
  analog electronic, 274, 291–294, 299, 300
    on/off action, 291
    PID action, 291–294

Controller *(continued)*
  defined, 5
  digital electronic, 274, 291, 298–300
    distributed control system (DCS), 299–300
    programmable logic controller (PLC), 298–299, 300
  electromagnetic, 274
  hydraulic, 274
  mechanical, 274
  pneumatic, 274, 291, 294–298, 299, 300
  PID, 297–298
  proportional, 294–296
Controller action, 274–291
  on/off, 274–277
    differential, 276–277
  proportional, 274, 277–291
Convection, 152, 156–158
  forced, 152, 156
  natural (free), 152, 156
Converter, 92, 135, 138–139, 141
  analog-to-digital, (A/D or ADC), 92, 138–139
  defined, 135
  digital-to-analog (D/A or DAC), 92, 138–139
  frequency-to-voltage, 141
  versus transducer, 135
  voltage-to-frequency, 141
Coulomb (C), 20–21, 55
  defined, 20
Couple, 256
Current (I), 2, 21–22, 24, 43, 44, 45, 51, 52, 53, 54, 55, 56, 58, 62, 63, 64–65, 66, 67, 68, 69, 70–71, 72, 73, 74, 84, 88, 89, 113, 130–131, 132–133, 134, 141, 165–166, 240. *See also* Electricity
  defined, 20
  effective, 69–70
  sinusoidal, 65, 69; *see also* Voltage, sinusoidal

D'Arsonval meter movement, 36–37
Dead-weight tester, *120*
Decibel (dB), 261–262
Deformation, elastic, 113, 114
Degrees, 148

Delta p (Δp). *See* Pressure difference
Density (ρ), 96, 152, 156, 186, 198, 217, 220, 244–249. *See also* Specific gravity
  measuring devices, 244–248
Derivative action, 282–284, 298
Devices, 2, 3, 5, 10, 80. *See also* Instrument
  active, 80
  density-measuring, 244–248
  differential pressure, 116
  displacement-measuring, 254–255
  electrical, 2, 80, 114, 236
  electromagnetic, 75
  flow-measuring, 210–221
    calibration of, 222
    installation of, 221–222
    selection of, 221
  force-measuring, 256–260
  humidity-measuring, 235–240
  hydraulic, 2, 104–105
  linear, 14
  magnetoelastic, 259
  mechanical, 2
  passive, 80
  pH-measuring, 265–266
    calibration of, 266
  pneumatic, 2
  pressure-sensing, 96, 105–117
    calibration of, 119–120
    installation of, 118–119
    protection of, 121–122
    selection of, 117–118
  sound-measuring, 262–263
    calibration of, 263
  temperature-measuring, 159–168
    calibration of, 171
    installation of, 170–171
    protection of, 171
    selection of, 168–170
  viscosity-measuring, 251–253
    calibration of, 253
Dew point, 232
  measuring devices, 238–239
Diaphragm, 35–36, 110, 111, *111*, 112, 113, 263
  applications of, 111
  materials for construction of, 111

Dielectric, 54, 114, 179, 236
Dielectric constant, 179, 184
Differential meter
   bell-type, 116, *117*
   bellows-type, 116, *116*
   weight–balance-type, 116–117
Differentiator, 83, 292
Digital communication, standards for, 138
Digital-to-analog converter (D/A or DAC), 92, 138–139
Direct current (DC). *See* Electricity, DC
Displacers, 183, 191–192
Display
   analog, 9, 12
   continuous, 9
   digital, 9, 12
Displayed value, 14
Distributed control system (DCS), 299–300
Drag
   form, 208–209
   frictional, 13
Dynamometers, 259

Electricity, 19–45, 51–75
   AC, 51–75
   DC, 19–45, 51, 60, 69, 70, 73
Electromagnetic field, 37
Electromagnetism, 70, 74–75, 152, 215–217, 274
Electromotive force. *See* Voltage
Electronics, 22, 24, 79–92
   analog, 80–90, 91, 92
   defined, 80
   digital, 80, 90–92
   division of, 80
Electrons, 20, 21–22, 53, 65
Elements, 5
Energy
   kinetic, 204
   potential, 204
Error signal, 279, 292, 293, 295
Expansion
   linear, 152, 158, 159, 160
   thermal, 152, 158–159
   volumetric, 152, 158–159, 160

Farad (F), 55
Feedback, 82

Fiber optics, 142
   defined, 142
Filters, 83
Flange, 210–211, 215
Floats, 182–183, 190–191, 214
Flow
   continuity equation, 201–202, 205–206
   current, 89, 164, 165
   formulas used in calculating, 201–209
   laminar, 198, 199, 218
   of liquids and gases, 2, 19, 96, 197–222
   measuring devices, 210–221
      calibration of, 222
      installation of, 221–222
      selection of, 221
   open-channel, 218–219
   total, 200
   turbulent, 198, 199
   types of measurement, 200
Flow nozzle, 210, 211, 212–213, 214, 221, 222
Flow patterns, 198–199, 210
Flow rate, 200, 201, 210–219, 251
   meters, 210–219, 222
Flume, 218–219, 221
   Parshall, 218–219
Force
   defined, 255
   measuring devices, 256–260
Frequency, 52, 53, 55, 62, 65, 67, 115, 141, 246, 261, 262
Frequency spectrum analysis, 262, 263
Friction, 199, 200, 206–207, 258
Frothing, 193

Gages
   draft; *see* Manometer, inclined
   ionization, 113
   McLeod, 113
   Pirani, 113
   pressure, 96, 97, 102, 110–112, 186, 193
   strain, *34*, 34–36, 114, 218, 259–260
   vacuum, 113
Gain, 84–85, 86–87, 90, 279, 289, 295

Gain *(continued)*
   coarse, 84–85, 86–87
   fine, 84–85, 86–87
Gaseous phase, 150, 152, 162
Gears, 30, 220
"Ground" voltage, 28

Head, 100, 203, 208, 218
Heat, 69, 70, 71, 72, 113, 134, 147–171, 218, 221, 232, 235
   amount of, 154–155
   formulas related to, 152, 154–159
Heat transfer, 151, 152, 157–158
Henrys (H), 62
Hertz (Hz), 52
Humidity, 2, 5, 19, 229–240
   absolute; *see* Humidity ratio
   defined, 240, 241
   measuring devices, 235–240
   relative (ϕ), 230–231, 232, 234, 235–236, 237
   specific; *see* Humidity ratio
Humidity ratio, 231
Hydraulic systems, 258–259
Hydrometer, 244–245
Hydrostatic paradox, 101
Hygrometer, 235–236
   gravimetric, 240
Hygroscopic, 236
Hysteresis, 13–14
   defined, 13

IEEE-488, 138
Impedance (Z), 53, 54, 64, 89–90
   defined, 53
In phase, 66
Indicated value, 14
Inductance, 61–63, 64, 68, 69, 114
   defined, 61
Inductive reactance, 62–63, 68
Inductor, 61, 62, 68, 80, 89
Infrared waves, 240
Instrument
   accuracy of, 8, 9–11
   analog, 2, 9
   defined, 2
   digital, 2, 9–10, 39, 40
   precision of, 8, 9, 11–12
   reproducibility of, 8, 9, 12–14
   static characteristics of, 8–15

Instrumentation, 1, 2, 14, 15, 20, 33, 34, 80, 88, 130, 131, 135, 140, 177
Insulator, 20, 54, 164
Integral action, 284–288, 298
Integrator, 83, 292

Kelvin scale, 149, 154
Kirchhoff's Laws, 42–45, 59
  current, 43–45
  voltage, 42–43, 44

Ladder logic, 298
Lag time, 274–275, 276, 290–291
  defined, 274–275
Law of intermediate metals, 166
Law of intermediate temperatures, 166
Length, 2
Level, 2, 19, 55, 177
  direct measuring devices, 179
  formulas for determining, 179
  indirect measuring devices, 179
Level measurement, 178, 184
  continuous, 178, 184
  direct, 180–185
  indirect, 186–190
  single-point (position), 178, 185
  solids, 194
  weight method, 188–190
Linear variable differential transformer (LVDT), 254–255
Linearity, 8, 14
  defined, 14
Lines of force, 62
Linkages, 30
Liquid phase, 150, 152, 162
Litmus paper, 265–266
Load cells, 35, 255, 259
Load change, 8, 293
Loudness, 262
Loudness levels, 262

Magnetic field, 61–62, 74–75, 217. See also Electromagnetism; Magnetism
Magnetic flux, 74
Magnetism, 14, 73, 74, 191. See also Electromagnetism; Magnetic field
Manipulated variables, 4, 5, 7, 20, 277

Manometer, 96, 105–110, 186
  fluids for, 110
  inclined, 109
  U-tube, 106–108
  well, 109
Mass, defined, 255
McLeod gage, 113
Measured variables, 4–5, 7, 137, 276, 277, 278, 279–281, 282, 283–284, 285, 286, 289, 290
Meniscus, 106
Meter, 263
  electromagnetic flow, 216–217
  flow rate, 210–219
  mass flow, 210, 220–221
  nutating disk (wobble), 220
  piston, 220
  positive-displacement, 220
  quantity, 220; see also Meter, total flow
  total flow, 210, 220
  turbine flow, 215–216, 221
  ultrasonic, 218
  variable-area, 214–215
  velocity, 220
  vortex precession, 217
Microprocessor, 80, 92
Microwaves, 239–240
Moisture content, 239–240
Modulus of elasticity, 260
Moment. See Torque
Multimeter, 2, 9

National Bureau of Standards. See National Institute of Standards and Technology
National Institute of Standards and Technology, 9, 240
Neutron reflection, 240
Newtonian fluids, 251
Noise, 260
  electrical, 134, 135, 141
  random, 261
Nonreproducibility, 13
Nuclear magnetic resonance, 240

Obstruction, 62
Offset, 134, 135, 295
Offset control, 85, 86
Ohm (Ω), 21
Ohmmeter, 2, 41

Ohm's Law, 22, 24, 31, 41, 52, 56, 65, 71
On/off control action, 274–277, 291
  differential, 276–277
Op-amp. See Amplifier, operational
Orifice plate, 210–211, 212–213, 214–215, 221–222
  concentric, 210
  eccentric, 210
  segmental, 210
Oscillation, 82
Oscilloscope, 2
Out of phase, 66
Overpressure, 118
Overshoot, 8, 289

Parabolic velocity distribution, 198
Pascal (Pa), 97–98
Pascal's Law, 104–105
Peltier effect, 165, 166
Percent reading (%Reading), 10
Percent span (%Span), 10
Percentage full-scale accuracy (%FS), 9, 10
Period, 51
Permeability, magnetic, 259
Permittivity (ε), 54–55
pH, 265–267
  measuring devices, 265–266
  calibration of, 266
Phase
  current, 66, 68
  voltage, 66, 68
Phase change, 52, 64–69, 150
Phase difference, 66, 71
Phons, 262
Physical variables, 2, 5, 14, 19, 30
PID. See Proportional control with derivative and integral action
Piezoresistivity, 259
Pirani gage, 113
Pitot-static tube, 213–214, 222
Pneumatic systems, 258, 259
Polarity, 43, 62, 82, 133
Position
  absolute measurement, 254
  incremental measurement, 254
Position/displacement, 41, 55, 254–255
  measuring devices, 254–255
Potentiometers (pots), 29–30, 89, 90, 254–255

Potentiometer *(continued)*
  defined, 29
Power (electrical), 69, 70–72, 73, 74, 82, 134–135, 291
  defined, 70
Precision, 8, 9, 11–12, 130
  defined, 11
Pressure (P), 2, 5, 19, 21, 41, 55, 95–122, 179, 180, 186–187, 200, 204–207, 210–211, 222, 246–247, 248, 258, 261, 294–296, 297–298
  absolute, 96–97, 204
  atmospheric, 96–97, 99, 101, 102, 112, 113, 148, 204, 234–235
  defined, 96
  dynamic, 101, 198, 204–205, 213
  fluid, 103, 104
  formulas for calculating, 99–105
  gage, 197, 100
  hydrostatic, 99, 186
  impact, 101
  measuring devices, 96, 105–117
  negative gage; see Vacuum
  proof, 118
  static, 100–101, 198, 204–205, 213
  total, 101
Pressure difference, 97, 106, 107, 210–214, 217, 221, 222, 246
Pressure differential. *See* Pressure difference
Pressure-sensing devices
  calibration of, 119–120
  installation of, 118–119
  protection of, 121–122
  selection of, 117–118
Probe, 184–185, 248, 290
  capacitance, 184
  conductivity, 184
  ultrasonic, 184–185
Process, defined, 2
Process control, 2–8, 14, 20, 130, 131, 133, 139, 273–300
  defined, 2
Processors, 298
Programmable logic controller (PLC), 298–299, 300
Proportional action, 274, 277–291
Proportional and derivative (PD) action, 282–284

Proportional and integral (PI) action, 284–288
Proportional band, 279, 281, 289–291
  adjustment of, 289–291
Proportional control with derivative and integral action (PID), 288, 291–294, 297–298
Proportionality constant (μ), 250. *See also* Viscosity
Psychrometer, 236–237
  sling, 237, *238*, 240
Psychrometric chart, 232, *233*, 234–235, 237
Pulses, 91–92, 136–137, 138, 140, 141
Pure tone, 261
Pyrometer, 167–168, *168*
  defined, 167

Radiation, 152, 157–158, 159, 167, 188, 193, 240, 247–248
Radioactive material, 188
Range, 170
Rankine scale, 149, 153–154
Rate action. *See* Derivative action
Reading, 14
Receiver, 6, 240, 261
Reflectance, 240
Relay, 75, 291, 297
Reluctance, 62, 114
Repeatability, 8, 14
  defined, 14
Reproducibility, 8, 9, 12–14
  defined, 12
Reset action. *See* Integral action
Resistance (R), 20, 21, 22, 27–28, 33–36, 40–41, 43, 52, 53, 54, 56, 61, 62, 63, 68, 69, 71, 72, 89, 114, 115, 134, 159, 164, 236, 251. *See also* Resistors
  defined, 20, 53
Resistive wire, 30
Resistivity, 41–42
Resistor coil, 30
Resistors, 20, 24–36, 43, 52, 55, 58, 61, 65, 66, 71, 80, 82, 132, 293, 294. *See also* Resistance
  color coding of, *25*, 25–26, *26*
  fixed, 30, 38, 39

Resistors *(continued)*
  in parallel, 31, 38, 56, 293, 294
  photoelectric, 239
  in series, 26–27, 28, 38, 56
  size of, 25
  types of, 24–25, *25*
  variable, 39
Resolution, 14
  defined, 14
Resonance, 193–194
Reynolds, Osborne, 198
Reynolds number (R), 198–199, 207, 208, 209
  intermediate region, 199
Rotameters, 214–215
RS-232, 138

Saturated conditions, 230
Schematic symbols
  electrical and electronic, *23*, 24, 33
  of operational amplifier, *82*
  of transformer, *73*
Seebeck effect, 165, 166
Semiconductors, 20, 80, 92
Sensor, 4, 5, 6, 8, 130, 138, 139, 167, 189, 215, 266, 278–279, 290, 291. *See also* Transducer
  defined, 5
  pressure, 110
    bellows, 110, *111*, 112, 113
    capsule, 110, *111*, 112
    diaphragm, 110, 111, *111*, 112, 113
    tube, 110–111, *111*, 113
  resistive, 34–36
Set point, 6, 7, 274, 275, 276, 277, 279, 280, 284, 288, 289–290, 295, 296, 298
  defined, 6
Shear stress (τ), 250
Shunt, 38
Sight glass, 180–182
Signal, inverted, 82–83
Signal conditioning, 134–135
Signal transmission, 8, 15, 20, 129–142, 218. *See also* Transmission
  electrical, 130–141
    analog, 130–135, 140
    digital, 130, 136–139, 140

Signal transmission *(continued)*
　telemetry, 130, 140–141
　fiber optics, 142
　pneumatic, 140
Sine wave, 52
Slider, 29–30
　defined, 29
Slope, 282–283, 287
Solenoid, 60, 75, 291
Solid phase, 150, 152
Sone, 262
Sound, 260–263
　measuring devices, 262–263
　　calibration of, 263
　primary source, 261
　secondary source, 261
Sound pressure level (SPL), 261–262
Span, 118, 134, 135, 168, 170, 295
　defined, 10
Specific gravity (SG), 100, 244, 249. *See also* Density
Specific heat, 151, 158
　defined, 151
Specific humidity. *See* Humidity ratio
Specific weight (γ), 99, 100, 103, 110, 122, 179, 186, 187, 190, 191, 244, 249
Spectrum analyzer, 263
Steady-state condition, 59
Strain gage, *34*, 34–36, 114, 218, 259–260
Sublimation, 151
Subtracters, 83, 86
Superconductors, 20

Taps
　flange, 210–211, 215
　pipe, 211, 221
　pressure, 210–211, 221–222
　vena contracta, 211, 221
Telemetering. *See* Telemetry
Telemetry, 130, 140–141
　defined, 140
Temperature, 2, 5, 7, 19, 33, 41–42, 96, 129, 131, 134, 147–171, 210, 218, 221, 222, 230, 235, 236–237, 238, 239, 245, 248, 251, 260, 266, 290
　ambient, 240

Temperature *(continued)*
　dry-bulb, 231, 232, 234, 237, 240
　effect on pH, 266
　formulas related to, 152–159
　measuring devices, 159–171
　　calibration of, 171
　　installation of, 170–171
　　protection of, 171
　　selection of, 168–170
　wet-bulb, 231–232, 234, 237, 240
Temperature scales, 148–149
　comparison of, 149
Thermal expansion, 152, 158–159. *See also* Expansion
Thermistors, 164, 170, 217, 239
Thermocouple, 6, 113, 131, 132, 134, 135, 164–166, 167
"Thermocouple laws," 166
Thermoelectric effect, 165, 166
Thermohydrometer, 245
Thermometer, 5, 231, 237
　bimetallic, 161, 170, 291
　defined, 159
　liquid, 160, 170
　mercury, 159–160, 170
　pressure-spring, 161–164, 237
　　Class 1: liquid-filled, 162, 170
　　Class 2: vapor pressure, 162–163, *163*
　　Class 3: gas-filled, 162, 163–164
　　Class 4: mercury-filled, 162, 164, 170, 291
　resistance, 164, 171, 237
　types of, 159–166
Thermopile, 167
Thomson effect, 165–166
Time constant, 58–61, 169, 293
　RC, 59–61
　thermal, 169
Time lag, 8
Titration, 240
Torque (moment), 194, 255–256, 257, 259, 294
Torr, 102
Transducer, 4, 5, 6, 7, 8, 96, 113, 114–115, 131–133, 138, 142, 218, 254, 256, 278–279, 295. *See also* Sensor
　capacitance, 114
　carbon pile, 114, 115
　current-to-pressure (I/P), 5, 142

Transducer *(continued)*
　current-to-voltage (I/E), 133
　defined, 5
　elastic, 257–258
　inductance, 114
　piezoelectric, 114–115
　pressure, 114, 118, 119, 187
　pressure-to-current (P/I), 142
　pressure-to-voltage (P/E), 142
　reluctance, 114
　resistance, 114
　resistive, 34–36
　voltage-to-current (E/I), 131–133
　voltage-to-pressure (E/P), 142
Transfer function. *See* Transfer ratio
Transfer ratio, 88, 132, 133
Transformer, 72–74
　schematic symbol, *73*
Transistor, 80
Transmission, 129–130. *See also* Signal transmission
　binary, 136–138
　defined, 129–130
　parallel, 138–139
　serial, 138–139
Transmitter, 6, 135, 184–185, 218, 240
　defined, 6
　torque, 255
Two-position action. *See* On/off control action

Vacuum, 101–102
　instruments, 113
Vaporization, 232, 236, 237
Variable signal
　manipulated, 129
　measured, 130, 131
Variables
　manipulated, 4, 5, 7, 20, 277
　measured, 4–5, 7, 137, 276, 277, 278, 279–281, 282, 283–284, 285, 286, 289, 290, 291–293
　physical, 2, 5, 14, 19, 30
Velocity, 22, 198, 200, 201–202, 204, 205, 207, 208, 213–214, 217, 220, 248
　defined, 198
Venturi tube, 210, 211, 212–213, 214, 219, 221

Vibration, 260
Viscometer, 251
   drag-type, 251, *253*
   inline falling-cylinder, 251, *252*
   Saybolt universal, 252
Viscosimeter. *See* Viscometer
Viscosity, 157–158, 198, 199, 210, 216, 217, 221, 249–253
   defined, 199
   measuring devices, 251–253
   calibration of, 253
Volt (v), 21
Voltage, 2, 7, 20, 21, 22, 24, 27, 28, 33, 36, 42–43, 52, 53, 54, 56, 59, 61–62, 63, 65–

Voltage *(continued)*
   66, 67, 68, 69, 70–71, 72, 73, 83–84, 85, 86, 88, 89, 91, 130, 131, 132, 133, 134, 135, 138, 141, 159, 166, 217, 254
   effective, 69–70
   sinusoidal, 68, 70; *see also* Current, sinusoidal
Voltage dividers, 27–30
   potentiometers, 29–30
   simple dividers, 27–29
Voltage drop, 27
Voltmeter, 2, 9, 33, 36, 38, 39, 69, 129

Volume, 22, 178, 244, 248. *See also* Expansion, volumetric
Vortex, 217–218
   defined, 217
   precession meter, 217

Watts (W), 70, 71, 72
Wavelengths, 240
Waves
   infrared, 240
   standing, 261
Weir, 218, 221
Well, still, 219
Wheatstone bridge, 33–34, 35, 36, 164